# LEITH-BUILT SHIPS

# LEITH-BUILT SHIPS

## VOLUME 1
## THEY ONCE WERE SHIPBUILDERS
## (1850 – 1918)

## R. O. NEISH

Whittles Publishing

Published by
Whittles Publishing Ltd.,
Dunbeath,
Caithness, KW6 6EG,
Scotland, UK
www.whittlespublishing.com

© 2019 R. O. Neish
ISBN 978-184995-443-3

# CONTENTS

# ACKNOWLEDGEMENTS

This series of books would not have been possible without the help of many people, some sadly no longer with us; if anyone is missed this is only down to my own forgetfulness and I apologise in advance. There are just too many to fit in here, but you know who you are and you have my thanks.

I have mentioned the loftsmen at the Henry Robb shipyard, who all helped to instil a sense of pride in ships and in the craft. I must also thank Robert Rowbottom once more, as he helped me more than he could have envisaged.

To all the men of the Leith Shipyard who influenced me in one way or another.

To Bob Sickles, ex-publisher of upstate New York, who encouraged me to pick up my writing again after I had all but given it up.

It would be remiss of me not to mention the ex-Robbs engineer J. Stevenson, who provided me with some initial information; B. Booth, shipwright, for many later photographs; and the archivists at Glasgow University for allowing me access to the wonderful Leith, Hull & Hamburg/Currie Line archive.

To the Scottish Records office for some initial help with information.

To Dr Stephen Gapps and his team at the Australian Maritime Museum.

To Dr William Collier and archivist Kathryn Preston from the oldest ship design house in the world, G.L. Watson Ltd., now based in Liverpool, who provided help, encouragement and some wonderful old original photographs.

To Lindsay Butterfield and all the other guys in New Zealand and Australia, who provided help and information – way too many people to list here.

To Peter McGee for some initial tidying up of my work, and my editor Anne Hamilton of WriteRight Editing Services at http://www.writerightediting.co.uk, who took the job on and helped me immensely in the editing/layout parts of my book before sending it to the publishers.

To Dr Keith Whittles and his team for believing in my project, and allowing me to bring this series of books to a wider public.

To Caroline Petherick at https://www.the-wordsmith.co.uk, for providing great direction and guidance in the completion of my book; she has been a pleasure to work with, while helping with some pretty complex and technical type details throughout this book.

To my darling wife Angie who has the patience of a saint, always believing in me and encouraging me to reach for my dreams, to never give up, and to persevere.

# FOREWORD

I spent over 30 years in Henry Robb's Victoria Shipyard, beginning as office boy, progressing through apprenticeship, ship's draughtsman and finally to naval architect, before the shipyard sadly closed in 1984. In my spare time I often strolled around Leith Docks, and my thoughts invariably wandered to what it must have been like in the early days, long before my time. I had heard tales of the København, the largest sailing ship ever built in Britain, from some of the old-timers in the drawing office. As you will read, she was built in Leith.

Years later I learned something of the history of the port of Leith, but not in any detail. Now in this wonderful book we can learn much about the early days of the shipyards and the many ships built there. Much of this was unknown to me, and it has been fascinating to read the wealth of information collated by Ron, a loftsman in the yard during some of the time I was there. Not only does he give valuable insight into the art of shipbuilding but he also pays tribute to the wonderful breed of men who worked on the berths – heavy work, often in the brutal winters of the Scottish east coast.

This book deserves to become a classic of its kind as it preserves for posterity Leith's proud shipbuilding history which could otherwise so easily be forgotten.

Robert W. Rowbottom MRINA,
April 2018, Edinburgh.

# PREFACE

As a shipbuilder, I started my working life as an apprentice loftsman, hoping to become a journeyman loftsman. An essential part of the trade was to have a thorough understanding of shipbuilding from start to finish – so is it any wonder that ships and the sea quickly got under my skin?

Shipbuilding in its purest form is an art, and it is skilled craftsmen who produce that art. Whilst science has a huge part to play, it has nothing to do with the emotion of building a ship. In my experience, there are few, if any, industries that could compare with shipbuilding for job satisfaction – pretty strange for work that is dirty and dangerous, with no health & safety to speak of, and virtually lacking facilities to help make the construction workers' day any easier. Working outside in all weathers, as did the shipbuilders in Leith, was considered normal – and unfortunately the better and faster you worked the sooner you might be out of a job.

Shipyards, by their very nature and location, became part of the social fabric and psyche of the local community. As such, it is almost incomprehensible to understand how this once great industry was allowed to go from building 80 per cent of all the mercantile ships in the world in 1893 to almost none less than 100 years later

I was very lucky to serve my time as a loftsman in the shipyard that was Henry Robb Limited at Leith. When I started my apprenticeship the loftsmen, all very skilled craftsmen, were (in no particular order) Peter Rennie, Jim Russell, Willie Weir, Ali Holland and John Conafray, along with the foreman, Bill Strawn. Each of them left many positives with me, but a special mention must go to Jim Russell, retired foreman loftsman. It was Jim who supplied me with the *Shipyard Build Book*, a full list of ships built at the Henry Robb Shipyard, from Yard No 1 to the last ship built, No 535. I must also mark the contributions made by the last naval architect to work in the Henry Robb Shipyard, Robert Rowbottom, who has been unstinting in his encouragement and information.

This, then, is the story of the ships built at the Leith shipyards. The first of three volumes, it is the story of some of the 1,150 ships built there. It covers the period circa 1850 to the end of the First World War, when the original shipyards, which were eventually to become amalgamated into the Henry Robb Shipyard, were still in existence. Volume II continues the story from 1918 to 1939.

* * *

To the many contributors to my website on the Leith shipyards – it would be impossible to mention everyone – thanks to all. Any opinion given is entirely my own, along with any mistakes, omissions or errors. I welcome corrections to the histories, and can be contacted through Whittles Publishing.

# INTRODUCTION

Much of the attention given to shipbuilding in Scotland has centred on the west, the Clyde in particular – and rightly so, given the number of ships built on this famous river. However, it is sometimes overlooked that the real centre of shipbuilding prior to the industrial revolution was in the east of Scotland. Before any shipyard of note was ever seen on the River Clyde, wooden ships had been built at Leith for several hundred years – a whole world of shipbuilding there stretches back through some 660 years of recorded history. From an article in the *Daily Mail* from 1937 on a visit to the Henry Robb Shipyard by a Glasgow journalist: 'the men of Leith were building ships when the men of Glasgow were sailing around in smacks poaching salmon.'

The Leith shipyards, so named for their base in the port of Leith, are the focus of this book, with particular reference to the history and ships originating from what would become the Henry Robb Shipyard. That story begins with the shipwright Thomas Morton, who started building ships at Leith in 1844. Morton was the man who had invented the patent slip, a means of raising ships from the water to the land on an even keel, thus enabling the vessel to be worked on without the expensive need to dry dock it.

Thirty-odd years later, in 1877, the company of Ramage & Ferguson Shipbuilders and Engineers was established; and after that, at the west pier of Leith Docks, was the shipyard of Cran & Somerville, which built ships from around 1880 until taken over by Henry Robb in 1927.

The next shipyard on the scene was that of Hawthorns & Co., which started on the banks of the Water of Leith prior to moving down to Shipbuilders' Row on the foreshore of Leith. Hawthorns took over the Morton shipyard in 1912 and continued until 1924, when it went into voluntary liquidation and was then taken over by Henry Robb. This acquisition gave the fledgling shipbuilders of Henry Robb access to three slipways, and meant that it could now launch ships directly into the sea.

All three of the above-mentioned shipyards, including the assimilated Morton's Yard, were to be incorporated into the firm of Henry Robb, ultimately to become Henry Robb Shipbuilders & Engineers Ltd.

*The Leith shipyards circa early 1930s with the yard of Ramage & Ferguson at the top and Henry Robb Shipyard (including J. Cran) below, the large white buildings. (Author's collection)*

In all, quite a line of pedigree shipbuilding and, indeed, some very fine vessels were built by the Leith shipyards in the 130-plus years that this history covers. While it would be out of the remit of a three-volume book to detail every ship, we shall endeavour to cover most of them: from the great days of sail, to the wonderful steam yachts, to the steamships of iron and steel.

But first, the port of Leith deserves some description in its own right. What was this place, close neighbour of Edinburgh, which become home to such an esteemed lineage of shipyards and shipbuilding? In fact, Leith, in the days prior to 1850, was a place that had a lot more in common with Glasgow than it ever had with Edinburgh, with which it was connected by a rough track called Leith Walk. Leith was an independent burgh until the mid-1920s when it was assimilated into the growing city of Edinburgh. This still rankles with some old Leithers.

Back then, the town of Leith was a rough, dirty place, one that you did not want to be strolling around. Times were hard, just as they were in most of Scotland, and the majority of Leithers were living in total slum conditions; some of the slums in the port were not demolished until the 1960s.

With a population of approximately 30,500, Leith was a growing town in the mid-18th century, but it had none of the infrastructure required to sustain its increase in growth – by 1880 the population had almost doubled, to around 55,000. The huge families were more a result of a survival instinct rather than a lack of television or birth control. As the infant mortality rate was drastic, a family needed to consist of many children just so that some of them would survive. Life expectancy was short in Scotland as a whole. If you made it much past 42, then you were doing well.

The man of the house had to have gainful employment. If he was from Leith, he would invariably be connected in some way with ships or shipbuilding. Any movement towards the decline of the industry meant he would have to move or emigrate to find somewhere for his talents to be put to good use. For the ordinary man who had been lucky enough to be indentured as an apprentice, it was a very tough life. At the end of a five-year apprenticeship, he would be welcomed into one of the trade bodies. But the old trade bodies are not to be confused with today's trade unions. No, this was more like a benevolent society of men who looked after each other, there being no other form of support for anyone who fell sick or had one of the many accidents which prevailed in this dangerous industry. When an apprentice qualified as a journeyman he became a freeman of his incorporation. On paying his entrance fee along with his quarterly dues, he would then be entitled to a small pension should he fall sick or meet with a bad accident.

The chances of falling sick were high in the Leith of the mid-19th century. The fact that hospital care – if available at all – was pretty rudimentary fostered a community of people who learned to look after their neighbours. As it was a port, many strange diseases were brought in by visiting ships, and of course the local ladies of the night also contributed to the spread of disease all around the town and further afield. Along with ex-servicemen coming back to the port at the end of the Crimean War in 1856 with many communicable diseases, it was not just a dangerous place to work but could be a dangerous place to live.

Things slowly began to improve by the turn of the century, when the Leith hospital was up and running, along with another called the Northern Leith Hospital. For the poor sailors (as the vast majority of them were) there was the Seaman's Mission, which performed sterling work to help seafarers through the years. This building is still in place, although it is now an upmarket hotel on the shore at Leith.

Schools were now being constructed, and there was finally a brand new purpose-built Nautical College. Opened in 1903, the college was the first such place in Scotland for the teaching of the skills required to go to sea and be a competent sailor or officer.

So not all was bad in Leith, but in the mid-19th century it must have been a foul-smelling place as there was no proper sanitation system in place – contributing again to the high incidence and spread of disease. The Water of Leith was a much-polluted river; by the time it got down to the seashore at Leith it had passed through Edinburgh on its meandering journey north. Edinburgh, with a far larger population – and this being just before the time of the Scottish Enlightenment – was also a smelly, dirty place for the most part. (Even today there are constant battles between the councillors of Edinburgh and Leith as to who should pay for the cleaning up of the old river, especially where the waste meets the containment barriers opposite the upmarket wine bars and eating houses which have replaced the traditional look of the Shore.)

Thankfully, the situation started to change as more and more piped water was allowed to go to the good citizens of Leith, making things a little cleaner and easier for them. As ever, the best way to deal with adversity is with humour and a sense of belonging when kindred spirits are formed, and this contributed to a community spirit which could not be broken nor downtrodden; many Leithers retain a strong sense of community identity to the present day.

There is no question that Leith, for all its contemporary gentrification, is a port town that owes its very existence to ships and shipbuilding.

# ONE: THE EARLY YEARS, 1850–1875

Leith is probably the most historic port in all of Scotland; records speak of shipbuilding there from the beginning of the 14th century. Records also indicate that once James I of Scotland had managed to secure his release from his 20-year imprisonment by the English, one of the first things he did was to establish a shipyard at Leith. This was in 1424, and accompanying the yard was a marine workshop where ships could be quickly adapted for warfare.

Since the vast sand bar at Leith prevented the building of very large ships, the largest ship of those times was built right next to Leith at Newhaven. The ship was the *Great Michael*, a carrack built in 1511 for King James IV of Scotland; it was said that it took all the forest of Fife to construct it. The *Great Michael* led to the creation of Britain's Royal Navy as we know it today, in that James IV was building a sizeable Scottish fleet which Henry VIII of England perceived as a threat. James being an ally of France, Henry had legitimate cause for concern, and proceeded to build up his own fleet, which became known as the Navy Royal. Amongst the new ships was the ill-fated *Mary Rose*, which has been reclaimed from her watery grave and now sits in Portsmouth. By the time Henry VIII died in 1547 his fleet had grown to some 58 ships. By 1660, a few more kings and queens of England having come and gone, Charles II of England inherited a permanent force of some 150 ships. It was a professional force, referred to as the Senior Service, and eventually it became the Royal Navy.

After the union of the crowns in 1603 (when James VI of Scotland also became James I of England) the first of the line warships, called HMS *Fury*, was built at Leith. There were also some large warships recorded as being built at Leith around the 17th century by the shipbuilders Sime & Rankin. That partnership built several warships in the days of the 'wooden walls'. Another first for Leith was the construction of Scotland's first dry dock in 1720, built close to Sandport Street. (We shall not mention the fact that the first real games of golf and the rules of the 'great game' originated at Leith Links!)

From one of the oldest shipyards in Leith was launched perhaps the best-known ship to be built in the old port. She was the steamship *Sirius* built in 1837 by the firm of Menzies

1

& Co. Ltd, the oldest of the shipbuilders in Leith. She was the first steamship to cross the Atlantic to the New World, from east to west – against the flow of the Gulf Stream – using steam power alone, arriving in New York a day before the famous *Great Western*. Menzies' yard and the *Sirius* were the beginnings of real steam-powered shipbuilding in the port. The building of the *Sirius* helped pave the way for many ship engine makers to take up the challenge of making engines for ships in Leith. The irony would come later, in that the steamship, with its ability to go almost anywhere at any time, would lead to the eventual demise of the busiest port in Scotland. Two large steamers were also built there and were said to be the largest afloat in 1840. There was doubtless exaggeration in the local news, but one of the steamers, RMS *Forth*, at 1,940 tons, was without doubt the largest ship built at Leith at that time.

The more permanent shipyards were based on the banks of the Water of Leith, the river flowing through the old port, which ran its relatively short course of some 35 miles into the wider Firth of Forth before it in turn opened out to the North Sea. This small river was never much more than 100 feet across at its widest, so it was remarkable how it sustained as many shipyards as it did. All the slips were angled downriver so that larger ships could be built. The furthest spot from the mouth of the river was occupied by the firm of Andersons until it was taken over by Hawthorns later in the same century; Andersons had built and launched *Gladstones*, one of the largest wooden ships built at Leith, in 1827. On the bank opposite Andersons stood the firm of Morton & Co. Restrictions to building larger iron ships led the firm to relocate to the foreshore of the dock area known as the western pier, which enabled ships to be launched directly into the sea. This would mean little or no limit on the size of the ship that could be launched in future.

Sime & Rankin, an active shipbuilding company around the same time, was responsible for the construction of many large sailing ships, along with the frigate HMS *Fox*, which saw action in the Crimean War. Prior to this, in 1826, Sime & Rankin had also built the large West Indiaman *Arcturus*. Another active shipbuilder around the Water of Leith and Leith Docks was the firm of Lachlan Rose & Son. It had a yard on the Water of Leith by the Sandport Bridge, near the firm of Robert Mackenzie & Co.

It was not only iron and steel ships that were built at Leith during this time; many fine sailing ships were also built and launched from the Leith shipyards. The very first steam-powered yacht was built and launched by Sime & Rankin in 1823. She was a wooden paddle steamer called *Quentin Durward*, and she, along with the launch of the *Sirius*, showcased the Leith shipyards as the forerunners of shipbuilding innovation at the time.

Although the capacity was there, by the mid-19th century times were changing fast. Shipbuilding in the west of Scotland was growing. Nor should the great shipbuilders of Aberdeen, to the north, be forgotten. In the days of the great clipper ships, they built some to equal any made in the west of Scotland or beyond. Perhaps the finest and fastest of all the great clipper ships built in Aberdeen was the magnificent *Thermopylae*. She was to have many battles with the better-known clipper *Cutty Sark*.

To paraphrase a piece in the *Daily Mail* of the 1930s:

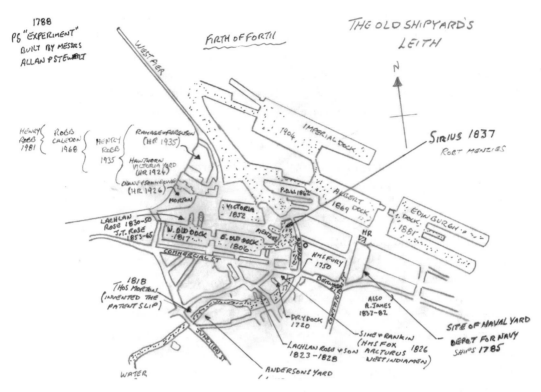

*Pencil sketch by R. Rowbottom, showing the approximate layout of
the main shipbuilders in Leith from the 17th century on.*

Within the recollection of many persons, shipbuilding was one of the
most important branches of industry carried on at Leith. About the same
time, other ships of the largest size were built at Leith, which led many
to suppose that the port would keep the lead in shipbuilding. During the
first half of the 18th century, Leith gave promise of being one of the
great shipbuilding centres of the country, but the Clyde seems to have
drawn the trade away from the port.

Industrial interest was prevalent in the west of Scotland. Although Edinburgh was content
to be involved with the more refined aspects – academia, arts and medicine – of 19th-century
living, industrial growth prospered in the west, as Glasgow grew and grew on the back of the
steam engine and shipbuilding.

In the days of sail alone, most of Scotland's trade had been with northern Europe and
the low countries of Holland, Belgium and northern France. But when Scotland, England,

Wales and Ireland were united, trade routes opened up dramatically. Glasgow, of course, was well positioned geographically to exploit this new business, which was to grow into an amazing industry, albeit built partly on some unpleasant parts of British history, such as war and slavery; the transportation of slaves to the New World was big business, as was the tobacco and sugar brought back on the return journey and tea from the East. Easy proximity to coal, and later to the new steelworks, meant the ships fed from the coal mines like big, black belching monsters.

*Drumrock, Yard No 108; the four-masted steel barque at anchor in an unidentified harbour. (ON 99316), 3182 tons, 329.2 feet × 45.4 feet ×25.7 feet. Built 1891 by Ramage & Ferguson, Leith. Owners Gillison and Chadwick, registered Liverpool. She subsequently became the German* Persimmon *and then* Helwig Vinnen, *later* Log Tye'e *and finally, having been converted to a barge in Vancouver, reverted to her original name. (State Library of South Australia PRG1373_6_49)*

As a result of Scotland's west coast's strengths, what should have been a great future for the shipbuilders at Leith stagnated for a few years as the money flowed west to fund the beginnings of the huge Clyde shipbuilding industry. While ships were still being built in the port of Leith, its main business became ship repair. Although this kept the Leith shipwrights busy in the winter and spring, when the ship owners had their vessels overhauled and improved, it did not provide continuous employment. So workers seeking regular employment would head to the north-east of Scotland or to the west. During this period, many wonderful clipper ships were built at those two centres of shipbuilding excellence. So until the end of the 19th century the port of Leith was often beset by problems for this and other reasons: if it was not those who owned the docks, it was a lack of capital to initiate the massive changes required to make the port more attractive to visiting ships – in particular newer steamships, which were becoming more and more prevalent. A huge dock improvement programme took over 40 years to complete, but once finished, the port could compete with most others in the British Isles.

In the shorter term, Leith rallied when the largest of its graving docks was completed. The Prince of Wales Dry Dock could at the time take any of the merchant ships in existence, with the exception of the *Great Eastern*. In addition, there were many other industries close by, all allied to the business of ships coming and going from the port. There was also a huge rope-making works in Leith, along with a thriving wine importing business. The rope works would of course begin to feel the pinch as the great sailing ships gradually yielded to the inevitable advance of steam power.

Another of the early shipping industries of Leith was the whaling industry. Whaling was a large employer in all the main ports on the east coast of Scotland during the 19th century. Whatever the rights or wrongs of that industry, it was a mainstay of the mariner's livelihood at the time, and there were several vessels involved in chasing the whale. They usually sailed from Leith on the same day of March, the destination usually being the fishing grounds off Greenland or in the Davis Straits. This involved a voyage of six to eight months. Occasionally the ship would also take in a trip down to the West Indies and while there take on a cargo of rum and sugar for the long voyage back to Leith, which at the time was one of the centres of the sugar-refining industry in Scotland. Later in that century, another name would move into the whaling industry, and this particular company would be around for more than 100 years. The name was to become a well-known Leith-based shipping company – Christian Salvesen & Co.

All around the British Isles were fertile fishing grounds, and by the 1850s in the Leith fishery district (which included Newhaven), there were more than 570 fishing boats worked by over 1,600 men and boys. Boats of all sizes took to the waters on a daily voyage to catch the great harvest of the sea, returning in time to land the catch for the markets. No matter what the weather was like these brave men went down to the sea, and many of them never came back.

Leith, of course, was also a seaport, and as such it depended on its continental trade. What was to become the Currie Line, together with several other shipping companies, operated out of the port on a regular basis. They traded to the Baltic ports and beyond, and ran regular passenger services between Leith and Newcastle. These short sea routes and coastal journeys

*A Fife fishing smack, typical of so many boats built from 1850 to 1890 at the many small yards at Leith. (Loftsman Collection)*

were the mainstay of the port, and other short sea passenger routes also opened up to the north – to Dundee and Aberdeen, some services continuing up to the Orkney and Shetland Isles. There was no bridge over the Forth at this time, which resulted in heavy sea traffic across the Firth of Forth, linking the Edinburgh and Leith side with the great coalfields of Fife. Then there were routes south to London. Thus, it was imperative that Leith had a docks and harbour layout that was conducive to this trade growing and thriving.

As steamships got larger, the voyage to North America and beyond became more accessible, turning into a trade that grew very large. Closer to home were the summer trade trips taken down the coast of the Firth of Forth, and many a trip was taken by the wealthy of the town by paddle steamer from Leith to towns such as North Berwick – which became a very upmarket tourist town. Trips could also be taken west from Leith to the smaller towns up the coast of the firth, such as Bo'ness and Grangemouth, and then to Alloa and beyond by horse-drawn carriage. Bo'ness had been a major port prior to Leith's emergence, but now became somewhat less of an important stop on the ship owners' calendar. Meanwhile the shipbuilders and citizens of Leith were more inclined to travel just a couple of miles along the coast and spend what little leisure time they had at the beach of Portobello.

With the port being more or less completely reliant on imports and exports, and receiving almost no assistance from Edinburgh – even though it owned the port as it had no access to the sea itself – the building of ships was high on the list of available employment. Shipbuilding, very hands-on work, required a fair number of skilled men to put a ship together. A multitude of assorted skilled trades were required as new methods and new materials were introduced to shipbuilding, and a great many builders and owners have, regretfully, gone unrecorded. A port's population could be transient, too; a small boatbuilder could set up on Leith sands above the high-water mark, build his boat and then move on.

However, with the advent of the 1848 California Gold Rush came a movement of humanity that was to tie up hundreds of the world's fast sailing ships for a good few years. With the Crimean War looming as well, both events would require many new ships to be built in Scotland as well as in North America.

# EARLY SHIPBUILDING AT THE VICTORIA SHIPYARDS

As mentioned above, two of the main shipyards that had been in the port of Leith were Menzies and Sime & Rankin. They were soon to be joined by a shipyard that would become famous even in its time. This shipbuilding venture had been started by the shipwright Thomas Morton, whose shipbuilding company, S. & H. Morton & Co. had moved from Bo'ness to Leith in 1844. It soon moved to the port of Leith, and set up on the land that was to eventually become the Henry Robb Victoria Shipyards. Thomas Morton had established a method of building ships and repairing them using his own patented invention, the aforementioned patent slipway.

> As with all the best inventions, it was as simple as it was ingenious. The patent slip was invented by Scotsman Thomas Morton (1781–1832) of Leith in 1818 as a cheaper alternative to a dry dock for ship repair. It consisted of an inclined plane, which extended well into the water, and a wooden cradle onto which a ship was floated. The ship was then attached to the cradle and hauled out of the water up the slip.
>
> *(Graces Guide)*

The problem with the patent slip lay in the fact that various cradles of a similar nature could also be built – as many were, particularly over in America.

It just needed a few changes here and there and – lo and behold! – you could claim it was of your own doing.

So, despite the fact he had been awarded the patent for his invention, many patent slips were built worldwide. There were a great many other variations of his design. To this day there are still slips in use that owe their use to the Thomas Morton-designed patent slip.

# TWO: IRON TO STEEL

Although the existing shipbuilding firm of S. & H. Morton & Co. now operated without Thomas Morton – he having died in December 1832 – it continued to build ships as part of a larger outlook on the general scene of the times. It was also heavily involved in the repair of ships, and was still building patent slips, but still it seemed to struggle along through some tough times in the middle of the 18th century. While the building of the actual ships did not constantly occupy the shipwrights of the firm, they were employed in the other branches of the trade, and were in high demand for annual maintenance. This was usually during the winter months – and there are not too many places of work colder than below a ship on blocks in a dry dock in winter time!

From 1844, until taken over by the shipbuilding firm of Hawthorns & Co. in 1912, S. & H. Morton occupied some of the site that was to become the Victoria Shipyards. Ships were being built there from the company's inception at Leith, although it did not build ships in great numbers and certainly not in comparison to the great strides happening in the west. The River Clyde saw shipyards springing up every other year at a time when all the ship owners were clamouring for steamships. This was largely due to the advent of the American Civil War in 1861 and the propensity to support the Confederate cause, which was overwhelming in Glasgow and the Clyde area. Many fast iron steamships were being built there under the orders of the owners who had huge tobacco interests in the American South; so it was not, perhaps, so much commitment to the cause as supporting the South and the continuation of slavery. Losses for the South would entail large loss of profit for the majority of the ship owners and builders who were entwined with each other on the Clyde. Not that the workers of the shipyards concerned themselves with who would use the ships; their job was simply to build them and build them well, which they did.

Five years of American Civil War meant full order books for some of the shipyards, since the vast majority of the ships used in it were built and financed by interests in Great Britain. Furthermore, a large proportion of those ships – more than 300 – were built on the Clyde, and others on the Mersey in Liverpool. The inherent quest for improvement in the building of faster and bigger ships gave these shipyards a natural technological lead over other yards.

The Union's naval blockade was designed – as all blockades are – to limit the opposing side's access to material and supplies; materials and supplies that were made available to the Confederates by the ships built on the Clyde for just that purpose. These ships, most of them side-wheel paddle steamships, were usually of a shallow draught and built for high speeds. These paddle-wheelers, driven by steam engines that burned smokeless anthracite coal, could make 17 knots. They were used to run the Union blockade and were thus known as blockade runners. Further large profits were to be made from them in the supply of munitions and other valuable materials for the Southern war effort, cotton and tobacco being the equally profitable return cargo.

The Civil War efforts in the Confederacy relied on the export of cotton shipped through the port of Mobile, Alabama, to Europe by way of Havana. Imports of armaments, powder, salt, medicines, blankets, iron, rope, machinery and other goods, including luxury items, did much to sustain the Confederates and support their morale. From January to August 1864, the city of Mobile saw more activity from blockade runners than any port in the Confederacy except Wilmington.

Because the American South lacked sufficient sailors, skippers and shipbuilding capability, the blockade runners were built, commanded and manned by British officers and sailors. On each trip a runner carried several hundred tons of compact, high-value cargo (cotton, turpentine and tobacco, outbound from the South; rifles, medicine, brandy, lingerie and coffee, inbound). Often they also carried mail. They charged from $300 to $1,000 per ton of cargo brought in, so two round trips a month would generate perhaps $250,000 in revenue (whereas $80,000 would be spent on wages and expenses). Private British investors spent around £50 million on the runners – a huge sum for the time, but the returns for the ship owners and investors were equally immense, and of course, with no physical risk to the wealthy armchair investor. Should you happen to pass some of the largest 19th-century houses on the outskirts of Glasgow or Liverpool, you might wonder how their original owners built and sustained such homes and staff while so many workers lived in slums. The answer is that the house owners might well have been ship owners, involved in one of the darker episodes of world history.

Ultimately, the cost in lost ships, and in the building or rebuilding of those lost ships, proved too much for the Southern states. In a way, these expenses contributed to the end of the American Civil War, in that the South could not afford to pay for any more ships to be built.

And so it was against this background that the company of S. & H. Morton continued to build ships at the Leith shipyards. It had other well-known shipping names around it, beginning with an amalgamation of the Hull & Leith Shipping Co., formed in 1880 and the Leith & Hamburg Shipping Co. (1816), to become known as the Hull & Leith Steam Packet Company Ltd. The new company started out by trading between the ports of Leith in Scotland and Hull in England. There was a further merger in 1847 with the Edinburgh & Dundee Steam Packet Co., the new company to be named the Forth & Clyde Shipping Co. The company changed names yet again, to become the Leith, Hull & Hamburg Steam

Packet Company Ltd when its ships began regular sailings between those three ports in 1852.

Ten years later, an already well-known local name in shipping joined the company – James (Jas) Currie. Helped by the fact his brother Donald owned the Castle Line and that many of the Leith, Hull & Hamburg ships were chartered to the Castle Line on the mail steamer route to South Africa, the company started to expand its passenger and cargo routes. In 1865 passenger and cargo routes were opened up to Copenhagen and Stettin, along with services to the north German ports, Russia, and the other main Baltic ports. This continued under the same company name up to and through the First World War. Further expansion followed until the year 1933 when it acquired the firm of M. Isaacs & Son of London, to gain a foothold in the Portuguese and Mediterranean routes.

As such, the company continued to trade until just after the start of the Second World War, at which point its name was deemed somewhat inappropriate for a country at war with Germany for the second time. (It had been in good company; the British royal family was mostly in fact German or of very close German descent, only changing its name from Saxe-Coburg to Windsor in 1917.) Thus, after something of a boardroom battle – eventually settled with common sense – the Currie Line Ltd was born, and proved to be a very good customer of the Leith shipyards in the early days.

Another well-known shipping line started off in Leith: Ben Line Steamers Limited, a Scottish shipping company pioneering the Far East–Europe trade. A private company, it was largely owned by members of the Thomson family from Leith and the Mitchell family from Alloa. The company had originated from a partnership between William and Alexander Thomson, who went into the shipping business in 1839. The partnership had started with sailing ships, importing marble from Italy and taking coal to Canada, returning with timber. From 1859 it operated routes to Singapore, China and Japan, and for a while this became its major source of business. By the mid-1880s, new routes were established to the Baltic, and the earlier Canadian trade was pared down.

The company changed its name to Ben Line Steamers Ltd in 1919. The Ben Line was one of the great shipping lines, and its distinctive style ships were known as the Leith Yachts by the men and women who sailed on them – but, alas, they are no more. For whatever reasons, the Ben Line did not order many ships from the local Leith shipyards after the First World War. This was perhaps due to its requiring larger ships than Leith had the capacity to build, and the Ben Line instead turned to Connell's yard at Scotstoun in Glasgow for a great many of its fine vessels.

Christian Salvesen & Co. was another company that established a fleet of steamships built and based at Leith. From 1880 its ships sailed from Leith to Stavanger in Norway, incorporating runs up and down the Norwegian coast before returning to Leith.

In 1883 owner Christian Salvesen delegated the management of the shipping operations to his eldest sons, Thomas and Frederick, who became partners in the firm. By the early 1900s, the company had developed significant interests in whaling, initially in the Arctic, and when this fishing ground had been almost cleared of whales its ships moved south, right down

*Salvesen harpoon gun on the Shore, Leith. The graffiti,
unfortunately, come with the times. (Author's collection)*

to the freezing seas of Antarctica. They operated from a base at Leith Harbour – named, of course, after home – on the island of South Georgia, to the south-east of the Falkland Islands.

Perhaps the oldest shipping line of them all, certainly in Leith, was the firm of Geo. Gibson & Son, or the (Geo.) Gibson Line as it was known to locals. But old or new, large or small, there were many more shipping lines in the port – and every order from every ship owner was very welcome to the yards in an increasingly tough market, as more and more shipyards opened up around the British Isles.

S. & H. Morton was one of the more prolific builders, frequently manufacturing ships for local ship owners of Leith, something that was often reflected in the ships' names: *Forth*, *Edinburgh*, *Leith* and *Mid-Lothian*, for example. The first ship built at the S. & H. Morton shipyard, at Yard No 1 (or Yard No 2 depending on which records are consulted), was the wood and iron paddle steamer *Forth*. She was a small vessel of 138 grt, just over 131 feet in length, and powered by a single-cylinder steam engine producing 70 hp and a single-ended coal-fired boiler, both built by the shipbuilder. The *Forth*, launched in May 1855, was an order from the Anstruther & Leith Steamship Company. She was to ply her cargo of grain and fish three times a week across the Firth of Forth from Anstruther to Leith, and this she did without mishap for 20 years. She was then sold back to the shipbuilder, Hugh Morton, who removed her engine and machinery, and leased her to the London & Edinburgh Shipping Company for use as a coal hulk. With her name unchanged, she served this purpose for another ten years before being sold for scrap in 1885.

Another early build from S. & H. Morton was named after Robert Napier who, ironically, is regarded as the father of Clyde shipbuilding (someone held in great esteem was called 'Faither' in deference to his superior knowledge rather than birth, relationship or age). This ship, at 435 grt, was launched in September 1844 (at Yard No 1 or Yard No 2, again dependent on record source) as the *Robert Napier*.

The ship called *Edinburgh* (ON 20432) was built in 1858 at Yard No 5. She was an iron screw steamer of 842 grt, and a bit longer than previous ships at just over 215 feet. She was powered by a two-cylinder steam engine with a single-ended coal boiler, both built and supplied by the shipbuilder. *Edinburgh* was launched in June 1858, an order from D.R. MacGregor, a wealthy ship owner who also went on to become M.P. for the Leith Burgh. (To become a member of parliament in those days you merely needed to have wealth.) He registered the vessel at Leith and used her for carrying general cargo across the often stormy North Sea. She sailed at regular intervals from Leith to Russia, and it was on such a voyage to the Russian city of Kronstadt, on Kotlin island 20 miles from St Petersburg in the Gulf of Finland, that the *Edinburgh* was mysteriously lost.

It was during the early winter of 1860 that she had left the port of Leith bound for the Baltic with a cargo of iron, paving slabs and cotton bales. With a crew of 24 and 6 passengers, the *Edinburgh* should have reached her destination around three days later – but she simply never arrived. About ten days later she was reported lost with all hands, and the search ships sent out to locate her found nothing. Reports of a terrible hurricane targeting Scotland and the Shetland Isles soon came to light, and it was suggested the route taken by the *Edinburgh* would have placed her in the path of the hurricane sometime in December of that year. A lifebuoy bearing the name *Edinburgh* was eventually found off the Fair Isles, but no trace of her captain or crew, along with a Dr MacKenzie and his family travelling as passengers, has ever been found.

The next interesting ship built at the S. & H. Morton yard was aptly named *Leith* (ON 43508). She was Yard No 9, another iron screw steamer at just over 245 feet in length, and another order from D.R. MacGregor. Launched in January 1868, she was also powered by a two-cylinder steam engine with a single-ended coal boiler, both of these built and supplied by the shipbuilder. The vessel was registered in Leith. Sadly, the *Leith* proved to be another somewhat unlucky ship for her owners, as she was to end up running aground on the island of Oesel, one of the many islands in the Baltic, off what is now Estonia. Just as her sister ship, the *Edinburgh*, had done, she had left the port of Leith with a general cargo bound for Kronstadt, but when almost at her destination she ran aground and was declared a total loss – only six months after her launch. It was becoming more obvious that this seaway from the east coast of Scotland was, as future sailors and airmen found out 90 years later, one of the most dangerous voyages undertaken, particularly during wintertime.

Sometimes ships were built on spec, not for any particular owner but simply to keep the shipyard going, and once the ship was completed the builder would hope to find a buyer. It was quite a common practice at the time, and it meant continued work for the men in the shipyard, giving some continuity to the builders. Without work, the men would have needed to go elsewhere. Such was the provenance of the iron screw steamer *Mid-Lothian* (ON 65773), which was built in 1871 as Yard No 20. She was a fair size of ship for her time, at 1,257 grt and with a length of just over 242 feet. She was powered by a two-cylinder engine producing 135 hp and fed by a single-ended coal-fired boiler, both of which were also built by the yard.

The *Mid-Lothian* was launched in November 1871 then fitted out, but she had to wait three years before a buyer was found for her. She was purchased by G.V. Turnbull and stayed registered at Leith. This ship was to go through a number of different owners (as is the way with many ships) before she met an untimely end in the First World War. She had the misfortune to be stopped by U-boat U-73 on 30 September 1917, just south of Cape Greco, Cyprus, and she was sent to the bottom of the Mediterranean by gunfire from the submarine on the surface.

Whilst some of the smaller boat builders were doing good business building fishing vessels, few wooden ships were being built around this time. It was more than likely that a ship which required a timber hull would be constructed via composite construction: her frames and internal structure built of iron, and her outside skin or hull sheathed with wood. Some ships were also sheathed below the waterline with a copper covering to prevent marine fouling. This arrangement of build meant a longer vessel with a higher cargo capacity, but all in all it was yet another compromise in the art of shipbuilding.

Ships built for speed would be of composite construction, while general cargo ships were built of iron. By this time, the methods for rolling flat iron plates had been discovered, and this brought forward the means of plating the whole of the hull with iron. Iron, though, was not an ideal material to work with in shipbuilding, being very labour-intensive and requiring heating to form the small plates into shape. However, it did offer tremendous strength, and also meant that larger and more economical steam engines could be mounted into the structure of these beamier vessels. The first all-iron-hulled ship had been launched in Glasgow in 1819. Called the *Vulcan*, she was to show the way forward in shipbuilding for the next 200 years, and a lot of innovation and technological advances had been made since her launch. And of course, the first practical steamship had also been built in Scotland; it had since as early as 1803 plied her way along the Forth & Clyde Canal. Iron ships, prior to the discovery of how to produce rolled plate, had their hull plates formed by heating, and the shape being pummelled into the plate. After that, all the rivet holes were marked into the plate, which involved much heavy manhandling of the iron. This was far more labour-intensive than the traditional wooden hull over a wooden backbone and ship's ribs which had been erected on the building blocks at the slipway of the shipyard. The old wood shipwrights would add that building iron ships involved far less skill …

It is certainly true that most shipwrights of the time wanted nothing to do with this new material, and this change sowed the seeds for the majority of demarcation disputes in the future. But iron and steam were the way of the future, changing shipbuilding for ever. The principles of the design of the first iron ship had, for example, transverse framing along with a double bottom arrangement and a centre line girder forming the backbone of the vessel. This has never really changed apart from the fact that a lot of ships are now also longitudinally framed (according to the Isherwood System, named after Joseph Isherwood, who managed to convince the shipbuilders of the day that this was a very good way of building ships). Just as with Brunel's *Great Eastern*, the same basic design principle remains in use today, with the use of longitudinal framing. Isherwood was by no means the only person to invent framing methods for the construction of ships, but he remains one of the better known.

The use of iron as a material for shipbuilding introduced a whole raft of new trades to the industry. Trades such as angle-smiths, ironworkers, riveters and their associated squads, along with the boilermakers, were a new challenge to the traditional domain of the shipwright.

From before recorded shipbuilding, it had been the shipwright who had built the ship. Only the master shipwright usually had the experience and keenness of eye to form the shape and arrange the build of a ship. In time, these skilled craftsmen became the shipyard managers, responsible for all aspects of the build of a wooden vessel. No plans or drawings were used in those days, and this created a sort of secret society of master shipwrights who were very reluctant to give up their long-held secrets. Some were better than others, and some ships, even though looking little different in their general style or amount of sails, were far better than others.

With the advent of iron in shipbuilding, the shipwright's position as king of the shipyard was being challenged and had to change. So it was negotiated that the shipwrights would take control of the erection and fairing of the steel or iron just as they had with the wooden ships. They would also control all aspects of the launching of a ship – a point that still causes confusion today, especially in non-British shipyards.

As the men in a shipyard always worked in squads, this meant that some of the shipwright branches also had to specialise in working only with wood. These highly skilled men worked just as the older traditional shipwrights had done when building ships; a specialised job that took a lot of training and was more akin to an art form than any other type of work to do with a ship. They became responsible for all wooden decks, hatch covers, and ship's boats – anything made of wood on the hull. Their remit, though, is not to be confused with the work of the ship's joiner, who worked on wood finishing inside the deck-houses and cabinetry.

The term 'ship's carpenter' actually referred to the shipwright, colloquially known as the chippy, who specialised in the wood aspects of steel or iron shipbuilding, and would be found on a ship at sea. These men worked with all the old traditional tools of the trade, wielding an adze with such skill as to form a straight and perfectly round tapering wooden mast from a single tree trunk the length of which you just won't see nowadays.

The shipwrights had a hierarchy that branched off in different directions, similar to that within any trade. The loftsman was top of the branch, the shipwright working with wood regarded as more highly skilled than his colleague working with steel or iron. Next came the shipwrights working on the slipway, those responsible for the ways and other ancillary work to do with the launch of the ship. The trades all worked a pyramid system, within which was the squad system, and this worked very well as each man became a specialist in some part of the trade. Should one man become a liability – not working as well or as hard as the others in the squad – he would be swiftly dealt with by the men themselves. It was a tough way to work in a tough industry, but on the other hand if one of the men had an injury or some other ailment (usually caused by an excess of alcohol) then the squad system would always help to carry him through.

The men who built the ships – the shipwrights, along with the boilermakers and the new trade of plater, and the riveters and caulkers and burners – formed what was known (in very non-politically correct terms) as the black squad. Their counterparts, responsible for the engineering and outfitting side of shipbuilding, were known as the outfitting squad.

Around the same time, the trade of naval architect started to be recognised as a legitimate branch of shipbuilding. This meant that the whole approach to building a ship structure out of iron had to be almost reinvented, and much work was done on the subject by some very eminent shipbuilders and mathematicians of the time; nowadays we might call them inventors. This was at the start of the industrial revolution and it was not just in shipbuilding that things were changing pretty quickly.

Wooden sailing ships had been around for hundreds of years, and while they still had a part to play for another couple of generations, the days of sail were on the way out. How the old traditional sailors must have felt on seeing what must have looked like some fire-breathing monster heading past them in the gloom of a rain storm with such consummate ease! The feelings of others might, though, have been echoed by Turner's 'The Fighting Temeraire'.

The introduction of iron as the material for shipbuilding dramatically changed the interests of the ship owners. It would also enable shipbuilders to build vessels of greater size than ever seen before, with the finest lines to allow them to go ever faster through the water and thereby increase the owner's profits.

So it was with the new material of iron that S. & H. Morton set about building ships at Leith and (along with the vessels mentioned previously) it manufactured the SS *Volunteer*, Yard No 7. She was built in 1861 for D. R. MacGregor, and was powered by a compound steam engine with two boilers driving power to her single screw. At 607 grt she was a large steamer for her times, and she was used on the routes from Leith to Holland. It was on this same route in February 1865, in conditions that were described as foggy, that she ran aground on the rocks off the coast at Gristhorpe, near the northern English port of Filey. Her captain and 22 crew had enough time to make it into the boats and they landed safely a few hours later, but SS *Volunteer* was a total loss after breaking her back on the rocks.

SS *Tom Morton* was built in 1872, Yard No 21, for Christian Salvesen of whaling fleet fame (the firm that successfully switched from the traditional Leith whaling grounds off Greenland to explore the Antarctic Ocean). The *Tom Morton* was a large iron cargo steamship of 1,402 grt, chartered during 1874. She took mail from Singapore while owned by H. Morton, and was then sold to G.V. Turnbull in 1876, but went missing in December 1886 on a voyage from Cardiff to Constantinople, and was never traced.

With the opening of the Suez Canal in 1869, the path was clear for enterprising ship owners to seek the riches of the Far East and the new markets of Australia and New Zealand – and the wealthy of Leith were no slouches when opportunity knocked. One of the men who moved into this new market was D.R. MacGregor, who had not only wealth but some vision.

With the aid of such local capital, ships and men were provided, and helped pioneer these new routes; they were to prove very profitable for the owners and captains, and guaranteed work for sailors and shipbuilders alike. That said, MacGregor himself did not have much luck, given that a large percentage of his ships ended up at the bottom of the sea.

When MacGregor's shipping company eventually went broke and was declared bankrupt in 1878, another well-known shipping family from Leith bought five of its ships. This company was that of Geo. Gibson, and his shipping line, along with that of Jas Currie, was to feature more than any of the other local shipping companies as a purchaser of local early Leith-built ships.

Shipyards that were now building iron ships powered by steam had either to turn their hand to making the engines for the ships, or buy their engines from some of the established engine makers – and by the 1870s these were all situated around the rapidly expanding city of Glasgow. The city was filling up rapidly with immigrant labour, most of which was attracted back across the Irish Sea from the Irish North – for many it was a homecoming. The huge steel mills being set up and operated not far from the Clyde shipyards were also employing thousands of workers; one of the largest of those yards employed in excess of 30,000 men. With easy access to steel plate the shipbuilders were of course going to use this new material, because it represented significant financial savings. Some of those shipyards had started out as engine makers, as was the case with J. Cran & Company of Leith, and the next logical step was to build the ship; all they needed was a suitable spot by the water's edge and a skilled workforce.

A number of shipyards turned to making their own engines, and in doing so classed themselves as shipbuilders & engineers. In fact, the three main shipbuilders of the waterfront at Leith Docks were all capable of making their own engines, which opened up another market, supplying to other shipbuilders. In hindsight, the steam engines of the period were rudimentary and somewhat Heath Robinson-looking contraptions but still very ingenious. The size of the engine was relative to the amount of power produced so they tended to be of vast size. Along with the coal-fired boilers (more commonly known as Scotch boilers), they took up a large area of space on a ship – valuable space which could otherwise have been used for cargo or passengers.

The size and power of the steam engine, along with the newly invented screw propeller, were to grow to levels which could not have been envisioned. Not everyone welcomed these advances, though; the staid controller of the Royal Navy swore, in the 1840s, that the placement of engines in his ships would never happen. Fortunately for Great Britain even this formidable force was eventually overcome, and in the years to come engines and screw propellers would power the ships of the Royal Navy to great effect.

But shipbuilders, too, continued to shy away from the new technology. It was only the forward thinking and vision of the great Isambard Kingdom Brunel that convinced the industry of the potential of the screw propeller. The massive steamship *Great Britain* was originally to be built as a side-paddle steamer, until Brunel managed to persuade the company ordering her that screw propellers were the way ahead. Slowly but surely the mode of traditional shipbuilding was changing.

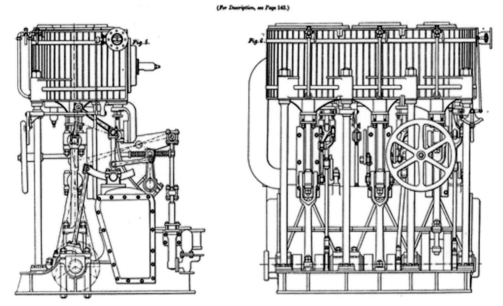

TRIPLE-EXPANSION ENGINES AND BOILER OF THE STEAM YACHT "GLADIATOR."

CONSTRUCTED BY MESSRS. RAMAGE AND FERGUSON, ENGINEERS, LEITH.

*(For Description, see Page 142.)*

*A typical triple-expansion engine, fitted to the steam yacht* Gladiator *built as ship No 74 at the Ramage & Ferguson yard in 1888. (From a sketch in an* Engineer & Shipbuilding *magazine, 1889)*

It was with the new triple-expansion steam engine that most ships would now be fitted. A better performing engine with lower fuel consumption, this would also lead the shipbuilders to consider putting two engines into a ship and so have her twin-screwed. This created a somewhat complex problem for the builders as they now had to fair a hull to accommodate two bosses as opposed to the single standard boss arrangement with one rudder. This new arrangement was to be designed into an R. & F. ship later. She was the twin-screw passenger ferry *Fatshan*. She was to be one of the first vessels to have this arrangement fitted successfully.

More in keeping with the new ideas of the industrial revolution, the iron ships were now truly iron: built with iron frames and no wooden support – no composite construction – and on the way to being manufactured in steel. The benefits of quicker construction time, added strength and an increase in the internal volume were unarguable. In addition, iron was also much cheaper to build with than wood, and did not require so many skilled men, which to the profit-conscious builder was an added bonus. Likewise, the apprenticeship was reduced from the six years it took to learn the basic skills of shaping and forming wood, to five years for working with iron and steel. The advent of iron as a shipbuilding material, coupled with the new and improved steam engines of the day, meant no longer did a captain have to rely on the vagaries of the wind. Rather, he could up steam and set off for whatever far-flung destination her owners had planned.

Nevertheless, change never comes quickly when a large group of human beings is involved, and it would still take some time to replace the trusted wooden ship known to shipbuilders and seamen for generations past. It was said that at one time a sailor plucked from the 17th or 18th century onto a sailing ship of the early 19th century would not feel out of place and would be able to function perfectly well within a few hours – that is how consistent the design and build of wooden ships had been. And there is no doubt that there were several design challenges with this new-fangled iron …

Although the material of construction had changed, it was still a tough form of work and the tradesmen were not called 'iron fighters' for nothing. At first, many seafarers exclaimed that it was impossible for an iron ship to float because, of course, iron sinks in water. Once this worry was defeated, another of the justified fears had been the magnetisation of the iron ship, in that it would render the sailors' main means of navigation unreliable. Once the industry understood that iron, and to a lesser degree, steel, is only magnetised by an electrical current being passed through it – becoming an electromagnet – and so a steel ship does not become magnetised (in the accepted sense) during construction, fears were reduced. However, it still remains that any large mass of iron will affect the accuracy of a magnetic compass, causing it to deviate wildly from magnetic north. And this problem was of course encountered when iron ships were first constructed. It was overcome by putting two large soft iron balls (Kelvin's balls) in the binnacle, on either side of the compass; these counteracted the effect of the hull, balancing the compass well enough to enable swinging to be carried out each year for the necessary fine-tuning.

Iron hulls did have a considerable drawback at the time as they were very susceptible to the collection and growth of marine life, which unless removed on a regular basis would eventually rot the hull beyond repair.

A bonus, though, was how the rigging on an iron ship was slightly different from that on the original wood-based vessels. The iron wire rigging made for a reduction in weight and less maintenance as it would last a lot longer than the traditional rope rigging. This in turn meant that masts did not need to be quite so thick or so high.

Meanwhile, although iron was replacing wood as the building material of the hull, wood was still used as deck material, and also for the small deck cabins commonly built.

Pros and cons aside, the development of iron ships simply became a necessity in the latter half of the 19th century. Due in part to the enormous potential for trade and profit, and the British imperial takeover of most of the known world, the shipping industry needed more speed – to reach (and deliver) valuable cargo and to return with it. The larger the ship, the more cargo she could bring back.

Thus it was technological, social, economic and world development factors that between them contributed to the demise of sail and the rise of the machine in the form of iron and steel ships powered by propellers driven by huge steam engines.

# THREE: SHIPBUILDERS' ROW

The last quarter of the 19th century was a time of tumultuous change during what was known as the industrial revolution. Abject poverty went hand in hand with great riches, and to the victors went the spoils – spoils rarely shared with the workers; the 'lower classes' were only there to provide the labour. Britain still ruled the waves, along with most of the known world, and it was said the sun never set on the British Empire. This had come as a result of the might of the Royal Navy, backed up by a huge merchant marine shipbuilding industry, which was now centred on four locations in the British Isles. The Clyde was one of them, along with the emerging massive shipyards in Belfast under the Harland & Wolff banner. In addition, shipyards on the three main rivers of the north-east of England – the Tyne, the Tees and the Wear – along with yards on the Mersey at Liverpool, would produce the vast majority of the ships built to keep this huge machine going. It is truly amazing to think that now, 120 years later, this industry is no more.

In 1875, things were quiet globally save a minor war in Afghanistan and another in South Africa. Locally, the first Edinburgh football derby was played between Hibernian and Hearts – Hibernian football club was to have a long association with the port of Leith and its shipyards. Still there were ships to be built; a good thing at a time in history when to be without a trade was a significant handicap. The only other viable option for an able-bodied working man or woman was to go into service and become a servant to the upper classes, or – for the men only at that time, of course – to join the army or the navy. Scotland, by percentage, provided an extraordinary amount of military resource to the government in London. It is often too easy to underestimate just how difficult these times were for members of the working classes, when life expectancy was, incredibly, only 29 years of age (this figure is all the more shocking since it does not take infant mortality into account). The industrial revolution had led to huge growth in many British towns and cities – Manchester, for example – but it was all built on the breaking backs of the working men, women and children. Things improved only marginally by the last quarter of the century.

At the start of the industrial revolution, all trade outside of the British Isles, and much within, was facilitated by ships. There were no aeroplanes as yet, and the steam train was just starting out, so the opportunities of striking it rich lay for the most part overseas. For the traders and ship owners the sea routes had been opened up by the decision to disband the charter through which the famous (or infamous, depending on your take on it) East India Company had ruled for years. Of course, it was in the interests of all ship owners to have the latest and fastest ships, which brought unimaginable wealth to the few in the position to exploit this wonderful new opportunity – and exploit it they did.

Now, what does one do with all this newfound great wealth when it comes along? In the 19th century, apart from large houses, expensive clothes and horses and a propensity for high-stakes gambling, there was not much the rich could buy to flaunt their status in society – no fancy cars or other gadgets – but there was the sea. And on the sea you could have your own ship powered by a brand new steam engine and designed by one of the better-known designers of the day. These vessels were called steam yachts: magnificent floating palaces that could take their owners anywhere, and were the ultimate status symbols. All the titled heads of state and more newly established industrialists coveted them. ('Steam yacht' was a title conjured up by a salesman in London trying to get more customers to use his boating service – but the name stuck.)

Some very fine steel-hulled auxiliary sailing ships for cargo carrying were also built in the last quarter of the century. Steam and sail power were both popular, but the owners and builders were tending more and more towards choosing the powered vessels. The large steel barques of more than 2,000 grt were still used on the long-haul run from Europe to the rich fields of nitrate in Chile; they became better known as the windjammers. With some famous captains amongst them, they were the stuff of a thousand novels. Battling around the notorious Cape Horn in a sleek, top-heavy sailing ship and making it home against all the odds, including some of the worst seas imaginable, the ships and crew fought for every inch of sea mile to beat their rivals. The cape was known as the graveyard of many a fine ship and the making of many a fine sailor.

Overall, it was the real golden boom time for shipbuilding in the British Isles, and it was no surprise that enterprising souls would be drawn to this work from all over.

> 'In 1862 the steam tonnage of the country was 537,000 tons; in 1872 it was 1,537,000 tons; and in 1882 it had reached 3,835,000 tons.' Chamberlain's speech, House of Commons, 19 May 1884.
>
> *(From Samuel Smiles, Men of Invention and Industry)*

# THE VICTORIA SHIPYARDS

So while the main focus of shipbuilding in the British Isles had shifted to the four main centres mentioned above, the shipbuilders in Leith were still occupied in the building and repair of ships, largely within the three building berths that formed the shipyard of S. &

H. Morton. Sailing ships continued to be manufactured there, but the primary business was in steamships.

During 1876 S. & H. Morton were involved in building a small steam-powered single-screw schooner launched as *Azalea*. She was of iron construction with a small engine giving her 20 hp, and at 62 feet in length and with her gross tonnage of around 35 tons she was a forerunner of the new trend in ships to be built. Then, not long after Morton's new neighbour R. & F. moved in alongside, the old firm was engaged in building two large ships, both of which went on to play a part in unfortunate accidents. These accidents, although many miles, and ten years, apart, had striking similarities.

The first one involved the *Esparto* (1,245 grt), launched on 20 September 1880. She was a single-screw steamer, schooner-rigged, and to be fitted with compound surface condensing engines producing approximately 140 hp. She was a fair-sized vessel at around 241 feet in length, with a beam of 33 feet and a moulded depth of 17 feet 3 inches, and was owned and operated at the port of Leith by the London and Edinburgh Shipping Company. The *Esparto* had had a useful working life of close to 19 years before, in November 1897, setting out on a voyage from Bo'ness, loaded with coal, and heading for Barcelona, Spain. With a crew of 19 and carrying a single passenger, she was well equipped with the modern life-saving appliances of the day – two lifeboats – along with a ship's boat and two compasses. She departed in moderate weather, proceeding south, and all went well until she arrived off Dungeness on the 27th and continued at full speed ahead into the early morning of the 28th. Another vessel was spotted too late, and this vessel, the *Noel*, rammed straight into the starboard side of the *Esparto*, almost cutting her in half.

The *Noel* was wedged into the *Esparto* for around four minutes, and some of the *Esparto* crew made to jump over to the other ship. This resulted in the death of two of the *Esparto's* crew as they fell overboard, and then, as the ship went down to her watery grave, she also took the only passenger on board with her.

The *Noel* was almost the exact same size and only 100 tons less in grt than the *Esparto*. She managed to limp on for a while before her captain realised that there was little hope for his ship. He made the decision to beach her two miles to the east of Dungeness, the weather so bad that the ship could not even communicate with the shore. Three days later she broke in two and became a total wreck.

The second ship to face tragedy was the *Iberia*. At 1,388 grt, and just over 254 feet in length she was another fair-sized steamer for her time. Launched in 1881, Yard No 30, she was a cargo vessel for her French owners, who registered her in the port of Marseilles. On 21 September 1888 she was on voyage from Basra, in the Persian Gulf, with a cargo of dates, wool and coffee, bound for New York. Meanwhile, the Cunard liner *Umbria* had set out from New York that morning, and in dense fog the two ships collided just off Long Island. While damage to the liner's bow was very slight, the smaller Iberia had around 14 feet sheared from her stern – mortal damage – although she stayed afloat for another 24 hours, allowing some of her cargo of wool to be salvaged. She sank the following day with some of the salvage party lucky to get off the ship when a watertight bulkhead gave way.

Not all of S. & H. Morton's steamships were any luckier. One of the largest to meet an untimely end was the *Scotia*, at 2,492 tons gross weight. She was built in 1881, as Yard No 31, being an iron screw steamer of 327 feet in length and powered by a compound steam engine and two single-ended boilers, also built by S. & H. Morton. Eight years later, while owned by the Boston Tow Boat Co and renamed *Mars*, she was wrecked near Los Roques Islands, Venezuela. Likewise, the *Wendouree*, Yard No 32, launched in 1882; a steel screw steamer (which meant that she was of steel construction with a single screw and powered by steam) with her dimensions given as 273.8 in length by 36.3 moulded beam by 19.3 moulded depth. She was powered by a triple-expansion steam engine producing approximately 230 hp. A standard coastal steamer of around 1,000 tons, she was built to trade around the coast of Australia; primarily her routes were from ports in New South Wales around to Port Adelaide in South Australia. But the *Wendouree* was wrecked in Australia in 1898.

This all serves to emphasise the fact that although ships were made of iron and powered by steam they were just as much at the mercy of the unpredictable ocean as any sailing ship ever was. One has to wonder whether this could have been caused in part at least by captains moving over from sail to the new steamships, as the control of these new ships did require a totally new and different approach; in addition a common result from official investigations revealed the continued use of many out-of-date navigation charts.

A much happier story was that of the *Falcon*, a ship that was to be seen around the Port Adelaide docks for many years. A small tug built by S. & H. Morton & Co. in 1884, her given dimensions were 96.9 feet in length by 19.3 feet moulded beam by 11.2 feet moulded depth. This iron screw steam tug had a long life; she was only broken up in Port Adelaide in 1961.

*Falcon, built by S. & H. Morton & Co in 1884, seen here in 1960 at Port Adelaide. (Author's collection)*

SS *Skulda*, Yard No 33, was launched in September 1888. This iron screw steamer was ordered by J.T. Salvesen & Co. of Grangemouth (not to be confused with Christian Salvesen of Leith) for use in its Baltic trade. She was a large ship, with her gross tonnage at 1,150, and dimensions of 225 feet in length with a beam of 32 feet 9 inches and a depth of 16 feet. She was to be supplied with engines to produce some 120 hp. She would unfortunately sink in a collision in the Firth of Forth in 1906.

The shipyard of S. & H. Morton continued to be busy with orders to fulfil (despite the rival R. & F. Shipbuilders moving in next door, to become an original 'noisy neighbour'), building ships such as the *Embla*, Yard No 34, of 1,172 grt. Built in 1882, she would not survive the First World War, hitting a mine en route from London to Dunkirk in 1915. SS *Embla* was a sister ship to the *Skulda*, also on order from J.T. Salvesen for trade in the Baltic. With almost the same dimensions as her sister ship, launched just three months previously, *Embla*'s gross tonnage was given at slightly more than *Skulda*, at 1,200, and her engine was supplied by Blair & Co. of Stockton, England.

Three more large ships built during 1883 were *Spider* at 1,181 grt, *Alvarado* at 1,031 grt and her sister ship *Pizarro* at 1,034 grt. Thus, Morton's Yard was both keeping busy and keeping up with the neighbours in terms of new orders. Morton also built a couple of paddle steamer ferries, the first launched in March 1883. She was PS *Lord Morton*, Yard No 36, fitted with a single diagonal engine, and had a length of 181 feet. This ship was not at her best in heavy weather as she tended to plough into the waves, something that was remedied when she was lengthened in 1900 by Hawthorns & Co., also at Leith. She survived the First World War and was sold to the Admiralty in 1918, finding herself in the White Sea where she was blown up to avoid capture by the Bolsheviks – quite a final voyage for this 35-year-old ferry intended only for inshore summer service in more peaceful times.

The next ferry was PS *Stirling Castle*, Yard No 40, which was launched in March 1884. She was a bit shorter than the *Lord Morton* at 160 feet in length, was again powered by a single diagonal engine, and she was to end up being sold to interests in Constantinople in 1898.

This had been a busy couple of years for the old shipyard, and it had also managed to build another two or three smaller ships in between the larger tonnage. The average build-time for a ship then was around three to six months, depending on the size of the vessel, the larger ones being on the stocks for about six months before launching. The iron screw tug *Condor*, Yard No 45, was launched at the S. & H. Morton shipyard in February 1885. She was built and designed for work on the River Thames, fitted with an improved sluice keel to help with turning, which by all accounts proved to be quite successful during her trials carried out on the same river on 12 March 1885.

The *Condor* was followed over the next two years by a number of smaller vessels of around the 140 grt mark, until the building of *Britannia* in 1885, given the yard number of 46. Now, here was a ship with an interesting working life, which included being sunk and then raised and reconditioned by the rival yard of Ramage & Ferguson. She was a steel screw steamer. Built by S. & H. Norton, she was launched on a cold day in February 1885 to go into the fleet of the Leith, Hull & Hamburg Steam Packet Co (J. Currie, Leith, as managers), she was registered

*SS Britannia Yard No 46, berthed in front of what was then the Seaman's Mission on the Water of Leith. (University of Glasgow Archives & Special Collections, Currie Line collection, GB 248 UGD 255/5/7/28b-britannia.)*

in the port of Leith. This steel vessel was 210 feet in length, with a beam of 27 feet and depth of hold 14 feet 6 inches, with a gross tonnage of 850 tons and powered by a compound steam engine with one single-ended boiler producing around 12 nhp (nominal horsepower).

After many voyages over some six years, sailing from Leith to the north of Germany, she was involved in a collision with SS *Bear* of Glasgow in January 1891, just off St Abb's Head in the Firth of Forth, not far from her home port of Leith. Although taken in tow by SS *Thame* she sank off Fidra island. It took until June of the same year before she was raised from the seabed in a salvage operation carried out by Ramage & Ferguson, which also went on to recondition her. The *Britannia* then went on to have another 24 useful working years before being wrecked on the Crumstane Rock, Farne Islands, while on a passage from Newcastle to Leith during the dark days of the First World War. She ran onto the rocks in September 1915, bringing to an end her 30 years of voyaging as a steel single-screw steamer.

The next ship on the stocks at the S. & H. Morton yard was the well-known and loved paddle steamer *Tantallion Castle*, launched in 1887 as Yard No 49.

| Friday | Lve Leith 9.30am, Portobello 10.10am, North Berwick 11.45am for St Andrews. Return from St Andrews 4pm, North Berwick 6pm, via Portobello to Leith |
| Saturday | Spare steamer (probably on roster 1) |
| Sunday | Lve Leith 11am, Portobello 11.45, Anstruther 1.30pm for St Andrews. Lve St Andrews 4.30pm, Anstruther 5.30pm via Portobello to Leith |

Builders S. & H. Morton, Leith 1887.
190' x 21.1' x 7.7'
Lost during First World War.

*An advertisement for the paddle steamer* Tantallion Castle. *(From an advertisement by the Galloway Sallon Steam Packet Co. (Author's collection)*

As the advertisement demonstrates, the steamer ran a circular route, starting out at Leith and going down the coast of the Firth of Forth to North Berwick before crossing the firth to the town of St Andrews, then voyaging back across the firth to return to Leith via Portobello – the well-known beach, now in the city of Edinburgh. PS *Tantallion Castle* was 190 feet in length with a beam of 21 feet, but as she seems to have had a problem similar to that of the PS *Lord Morton* she too was lengthened at Leith in 1895. She was the flagship of the fleet of Forth Ferries, but later went the same way as PS *Stirling Castle*, being sold to Constantinople in the same year, 1898. The *Tantallion Castle* was another vessel lost during the ravages of the First World War.

*A postcard showing the side-wheel paddle ferry* Tantallion Castle, *Yard No 49.*

Another two ships at more than 1,000 grt were in progress at Morton's during the time that R. & F. moved in to set up its shipyard next door. The wonderfully named SS *Kopernikus*, Yard No 53, was launched on 30 March 1889; she was 180 feet in length with a beam of 28 feet and depth of 15 feet, powered by triple compound engines. She had been ordered by Marcus Cohn & Son of Konigsberg, to be employed in the Baltic trade.

The *Kopernikus* was quickly followed by the build of a further three ships. The *Norna* was 1,144 grt and she was launched in July 1889; she had been ordered by J.T. Salvesen & Co. of Grangemouth and was for use in its Baltic trading routes. She was 232 feet in length and 33 feet in the beam with a depth of 16 feet, and powered by triple-expansion engines fitted by the builder; she was expected to be completed after successful sea trials the following month. Her engines, built by S. & H. Morton, exceeded the required service speed while on trials on the Firth of Forth, achieving a speed of 11.7 knots.

SS *Mabel*, at 376 grt, Yard No 55, was launched on 10 August 1889. She was 160 feet in length with a beam of 23 feet and a depth of 12 feet 6 inches, and was designed and built specially for trade between London and Paris. SS *Capella*, Yard No 56, was another Baltic trader, on order from Holm & Molzen, Flensburg. She was 223 feet in length with a beam of 32 feet 9 inches and 16 feet in depth, fitted with the builder's triple-expansion engines, and she was launched into the Forth on 5 November 1889. The *Capella* was followed the next month by the launch of SS *Jarnac*, Yard No 57, going into the water on 21 December 1889. She was an order from T. & J. Harrison of Liverpool, and intended specially for the wine trade between Liverpool and France. She was 185 feet in length, with a beam of 28 feet and depth of 14 feet 6 inches.

All in all, 1889 had been a very good year for the shipbuilders at S. & H. Morton, who also continued with a lot of ship repair work. The firm continued into 1890, one of the steamships built being SS *Otra*, Yard No 58, at 788 grt. She was ordered by Christian Salvesen & Co. of Leith for Norwegian owners and was 190 feet in length with a beam of 30 feet and depth of 15 feet 9 inches. With a large cargo capacity and good accommodation for passengers, she was launched on 19 February 1890, with her triple-expansion engine fitted by the builders. She was to work for some years before falling to a German U-boat in 1916; scuttling ships by opening their sea-cocks was a favourite pastime of the German submarines at the time. Renamed SS *Cap Mazagan* under French owners, she was stopped by a German U-boat (UB-38) on 1 October 1916 on a voyage from Port Talbot in Wales to France with a cargo of coal. The crew were ordered off the ship into the ship's boats and SS *Cap Mazagan* was sunk.

By this time the neighbouring shipyard of R. & F. Shipbuilders was well up and running; there must have been a bit of natural animosity between the two yards, and there was also another small builder of trawlers, D. Allan, right next door to R. & F. Although by no means a threat to the larger two yards, Allan would still be competing for labour. With three shipyards together there would always be the tendency between the men to brag about who built the best and who built the best the quickest. Fortunately for the shipyard owners there was more than enough unemployment in Leith to ensure that a steady stream of skilled men would always be available. Once in work they were not too difficult for the managers to deal with, as the threat of quickly being replaced was, of course, always there; this was a ploy used by management throughout the ensuing 100 years of the industry. Indeed, the practice still goes on today, though to a somewhat lesser extent.

# R. & F. SHIPBUILDERS AND ENGINEERS

The shipyard of Ramage & Ferguson started out in May 1877, established by two former managers of the famous Denny Bros shipyard at Dumbarton on the lower reaches of the River Clyde.

Richard Ramage had been the yard manager at Denny's well-respected shipyard at Dumbarton from 1870. His new business partner, John Ferguson, after an apprenticeship with Barclay Curle, had also worked at Denny. (NB Henry Robb had also been a manager at the famous shipyard at Dumbarton.) Ramage and Ferguson decided to branch out on their own, and soon saw the ideal waterfront land, known as the Victoria Shipyards, at Leith.

The land being owned by the Leith Docks authorities, it took the two men almost three years of negotiations with the Dock Board before they were allowed to lease part of the land to form the shipyard which was to build some of the finest vessels that ever graced the water. The Leith Dock Commission could apparently see only the fact that it could get more from the land by leasing it for the storage of imported timber. (It would seem that even in more recent times, under the guise of the Forth Ports Authority, its attitude indicated that it had rejected the idea that a shipyard on the site might bring some much-needed skilled employment to the port of Leith for the next 100 or so years.) Even once Ramage and Ferguson's negotiations were concluded the deal almost broke down again due to the many restrictions that Leith Dock Commission was trying to place on the new shipyard.

To give some due to the Leith Dock Commission, there may well have been some objections from the shipyard of S. & H. Morton. After all, who wants a direct competitor to set up next door? Even if their shipbuilding markets were a little different, the two yards would inevitably on occasion be chasing the same orders.

Fortunately, however, some common sense seems to have prevailed and the new shipyard was put in place. The new Ramage & Ferguson Shipbuilders and Engineers soon had its first order and in the late months of 1877 its workforce started building the first ship, to be launched in 1878. SS *Shamrock* was an iron-hulled ship of 591 grt, powered by a compound steam engine, and built for Crawford & Co. Ltd. She was to give good service before being sold on in 1897 to her new owners, who decided to have her lengthened in 1904. Her tonnage went from the original of 591 registered gross tonnage up to 762 grt. This first ever ship built by R. & F. at Leith would unfortunately not survive the First World War, meeting her demise in May 1916.

While the *Shamrock* was the first of the iron ships to be built by the new shipyard, it would also build some auxiliary sailing vessels. The famous tea clippers were manufactured via composite construction (iron frames and timber outer materials) in order to maximise their speed –critical for the owners, to maximise their profit margin. Furthermore, ships were still built in the old wooden traditional way by 'horning up' the keel and stem bar on the stocks, with the aid of many guy wires. With the shipwright's keen eye for fairness, the main central strengthening member, the keel, and centre line girder were erected on the keel blocks which formed the building berth. The steelwork was all set at an angle to the world of approximately 4 degrees, which allowed the ship to slide into the water when released from the standing ways.

By the time R. & F. was established, the preferred material for building steamships was with iron plates. Each plate had to be individually formed, and then, using a template supplied by the loft, all the rivet holes would be marked for punching out. Only then could the plate be hoisted up onto the skeleton frame of the ship on the slipway, lined up and drifted into place with bolts holding the plate in place; it was then ready for the team of riveters to do their job. Many a finger was lost by the natural reaction of placing a finger into the hole only for the plate to slip. Tales were told of this happening to men who, after a brief visit to the local hospital to get the wound cauterised and covered, were back on the job in hours.

Riveters worked in squads of four or five. There was the heater boy who – just as it sounds – heated up the rivets, all in sequence, and picked out the required rivet by the sound that the hot metal made with his tongs. He would then throw it, with skill that would shame a cricketer or pitcher, some 30–40 feet in the air up to the catcher who immediately caught the red-hot piece of metal in his shaped bucket. Next, the placer would place the rivet in the correct hole for the backer, who was inside the shell plate. The backer would work in tandem with the knocker, who was on the outside of the shell plate, to work at pace with alternate strikes of their 5lb hammers to form the rivet in the shell plate. They had around 30 seconds to work the rivet before it became too cool to be worked, before moving on to the next hole (and so it would continue all shift.) The cooled rivet was then swabbed with paraffin – and if there was the slightest sign of any of the liquid the faulty rivet would need to be removed, costing the swearing and sweating squad time and money, which was difficult enough to earn anyway. They were paid by piecework, meaning that they got paid by how many rivets they managed to complete in a 10-hour shift, which along with 4 to 6 hours working on a Saturday, formed the normal working week of around 54 to 56 hours. A good rivet squad would get through about 87 to 90 rivets an hour, and at the end of the day the hated counter would come out of his warm little office and put a chalk tick mark across each one.

As the building of ships in Leith got under way with the new firm of R. & F. Shipbuilders and Engineers, another shipbuilder set up adjacent to it. This was the aforementioned D. Allan, who carried on business there and at Granton, and was perhaps best known for being the first company to build a dedicated steam-powered trawler in Scotland. She was named *Pioneer* and she was a wooden-hulled vessel with a steam engine along with two masts and complete with sails. D. Allan & Co. specialised in the building of vessels to supply the huge fishing fleets around the British Isles. The company launched the stern trawler *Kingfisher* in September 1879, with her trials concluded in the Forth the following month. She was 75 feet in length with a beam of 16 feet and depth of 8 feet 4 inches. The steam trawler *Mercader* followed in April 1880, for Mercader & Son of San Sebastian, Spain.

Between May 1880 and February 1881, the firm continued building what it was very good at – its steam trawlers – along with another small steamer for use by the French Naval Authority between the port of Toulon and the island of Hyères. This vessel was named *Ocean Rover*, built for an owner in Granton, just past the harbour of Newhaven, just along the coast from Leith. It was not long after this launch that D. Allan & Co. moved out of the foreshore that would become Shipbuilders' Row, back to Granton; but shortly afterwards the company went into liquidation. Subsequently, R. & F. Shipbuilders took over Allan's vacant slipway in Shipbuilders' Row, giving it four slips for building and launching.

The new firm of R. & F. Shipbuilders had, it seemed, moved into Leith at the right time, there being a huge demand for new ships in this industrial age. It was a time of reasonable peace, too, as although the British were heavily involved in quelling what they then called the Indian Mutiny, the British Isles had not seen a major war for a wee while.

The *Shamrock*, described above, was quickly followed by Yard No 2 and Yard No 3 – following tradition, they were given the yard order number until they were christened with their name, painted and gleaming on the side of the newly plated and painted hull. Yard No 2, the first in a long line of steam yachts to be built at the Leith yard, was named SY *Ranee*, an order for a well-known local yachtsman, Thomas Glover.

Part of the deal struck with the yard was for Glover to take the *Ranee* around the ports of the British Isles and show her off to prospective new customers of these magnificent new vessels. She was launched into the cold waters of the Firth of Forth on 12 January 1878. This fine iron steam yacht was to eventually end her days in China, named *Pin Seng*, and she was deleted from Lloyd's Register in 1924.

*Albatross*, a screw steamer, was to be Yard No 3, and *Zephyr*, Yard No 4. She was another screw steamer, and they were launched within months of each other; four ships in three months following the *Shamrock* sliding down the ways into the basin of Leith in 1878. Then the small iron-hulled fishing craft, the *Stornoway*, followed them all. It must have been a real spur to the shipwrights of the two rival larger yards to look over into each other's shipyard and see each vessel taking shape on the adjacent slipways. They would also be able to measure how work was being done by the progress of the plating erected on each ship.

SS *Albatross*, Yard No 3, had an amazing longevity; she was broken up in 1994 after some 116 years of working, and that was only after a collision in the Bosphorus just off Istanbul.

Had she not been involved in that collision, who knows how much longer she could have gone on working and making money?

I would question if any ships being built today will still be around in 100 years' time, let alone still working; *Albatross* was an outstanding example of the craftsmanship of the shipbuilders of Leith.

In its first full year of operation, R. & F. built ten ships, giving an average of a ship built to launch stage every five or six weeks. By its third year of operation, this had risen to a total of seven ships built during each preceding year, working out at around one ship built to launch stage every seven and a half weeks or so; the shipyard was doing pretty well. All this at a time in 1879 when shipbuilding was going through one of its many down cycles; so few were the orders that many shipyards on the upper reaches of the Clyde had been closed down for a time. The forward-looking firm of William Denny, on the lower reaches of the Clyde at Dumbarton, was the exception, with a healthy order book.

But even with the downturn in shipbuilding the new company of R. & F. Shipbuilders was doing relatively well, with an order book and launches over the year amounting to a respectable seven ships, figures that would have today's shipyard managers salivating. Many of R. & F.'s ships were single-screw steamers with a schooner rig for sails – being three-masted as well, they must have made an impressive sight when launched at Leith. The launch of a ship was still a big deal in the days before radio or television, and most launches would draw a large crowd out to watch, flags and cheering creating a kind of holiday occasion for the watching crowd. Less so for the men who were building the ships, as once one was launched it was straight into building another one while the orders were there.

R. & F.'s next ship was the *Titania*, Yard No 13, an iron single-screw schooner-rigged steam yacht ordered by Mr John Lysaght, and then to be owned by the Marquess of Ailsa. She came alongside the *Vanadis*, Yard No 16, another iron single-screw schooner-rig steam yacht. The *Fair Geraldine* was a three-masted iron auxiliary schooner, built and launched in 1880, as Yard No 20. Her sailing rig, like that of the *Vanadis*, was designed in house, this time for an order by Harry L.B. MacAlmont. She was later renamed *Geralda*, and she was classed by Lloyd's as a steam yacht.

R. & F. Shipbuilders built a total of five steam yachts in its first three years of operation.

In between the steam yachts, R. & F. built two general cargo ships and a dredger for the Leith Dock Commission. The first of the cargo steamers was SS *Aeolus*, Yard No 14, at 497 grt, launched in November 1879, and the next, the *George Gowland*, Yard No 15, launched in January 1880. The firm then received an order from one of the oldest shipping companies in the port of Leith. This was Geo. Gibson, requesting a steam cargo ship of around 593 grt. She was launched in May 1880, named SS *Woodstock*. She had engines of 85 hp, built by Fleming & Ferguson of Paisley.

The iron screw steamer SS *Starley Hall*, Yard No 21, was then launched on 16 October 1880 at Leith, having been built for the London & Edinburgh Shipping Company at 610 grt, and powered once more with engines supplied by Fleming & Ferguson of Paisley. On her sea trials she exceeded her service speed and was timed over the measured mile at 11 knots.

In fact, so good were her sea trial results that as soon as she returned she was at once placed on the loading berth for London. SS *Starley Hall* would survive until broken up in 1928.

On the other hand, *Craigrownie*, Yard No 24, had a very short working life: just a couple of months. She was launched on 22 September 1881, an iron steamship of around 879 grt, a sister ship to *Craigallion* (of which, more in Chapter 4). But after just two voyages a storm drove her onto the coast of Ireland and she was wrecked on North Rock, Co. Down, on 13 December 1881.

The tenth and final ship to be launched in 1880 went down the slips on 2 December. She was Yard No 25 and was named *Wycliffe* on her launch, an order for a new line that had begun in the Far East for the Wycliffe Steamship Company Ltd. At 1,047 grt, she was a large ship for the yard, and her launch completed another pretty successful year for the firm. SS *Wycliffe* was to trade successfully in the Far East before she was blown up while at anchor in 1895 at Kinchow, Manchuria, during the Sino-Japanese war of 1894–1895.

The year of 1881 began with the launch of the first of four steamships to be used in the China trade. The first was launched on 15 January and named SS *Kama*, Yard No 23, though her name was quickly changed by her new owners to SS *Eaglescliffe* – she would actually go through a few name changes up until 1916. She had been built on spec by the builder, confident – rightly – that it could find a buyer for this large steamship of some 871 grt. During the First World War, while owned by a Norwegian shipping company, she was renamed *Dania*, and on 26 September was stopped and sunk by shellfire from UB-43 some 7 miles north-east of Nordkyn, mainland Europe's most northerly point. She was on route to Leith from Onega, on the White Sea near Archangel in northern Russia. Fortunately there was no loss of life.

Twelve days after the *Eaglescliffe*, on 27 January 1881, SS *Penang* was also launched. After that came SS *Ranee*, Yard No 27, launched in March 1881. Both the *Ranee* and the *Penang* were passenger/cargo vessels of approximately 625 grt each. SS *Ranee* was an order for the Borneo Company Ltd of Singapore and designed for work in the tropics, all her woodwork being of teak. Powered by engines producing some 105 hp, she exceeded her required service speed while on sea trials in the Firth of Forth, achieving a speed of 11½ knots. She was duly handed over to her delighted new owners, and promptly set out for the Far East to begin trading between Singapore, Sarawak and Borneo.

Meanwhile, SS *Penang* was an order from John Warrack & Co., Leith, for Chinese owners based in Singapore. Built with a large capacity for the carriage of rice, she had a shade-deck which (I quote from *Engineers & Naval Architect* magazine of the time) 'offers shelter to a large number of native passengers, while in the poop and deck-houses first class passengers and the European Officers have ample accommodation'. SS *Penang* would survive until fire took her, 23 nautical miles off Singapore in 1926, while SS *Ranee* was lost in China and deleted from Lloyd's Register in 1924.

R. & F. Shipbuilders also built some large sailing ships – barques, to give them their correct name – during its first ten years or so of operation, and, as we have noted above, from 1880 it had started to build ships using steel as the primary construction material, as it had the advantage over iron by being easier material to fabricate and could be forged, while still keeping all the strengths of iron. This meant that the iron workers were now plate workers – 'platers' as they became known. When the inventor, Henry Bessemer, revealed his revolutionary new way of turning iron into steel in England in 1856, it was to have a huge and dramatic imprint on the world of building vessels and vehicles to transport people and goods around the world. Iron remained in use for at least another 25 years, but steel had arrived as the main material of construction for the building of ships, just as it still is today. As ever, the shipbuilding industry took to the new material slowly but surely.

By 1858 the first all-steel ship in the world had been built in England for the great Scottish explorer of Africa, Dr David Livingstone. The *Ma Roberts* was to help with his exploration of the Zambezi River in his quest to find the sources of the Nile. Although the ship was not a success, her poor-quality steel being unsuitable for the hot and humid conditions the explorer would meet, with better quality steel becoming available she led the way for the use of this material.

Steel materials meant that the Leith shipyards could also build much larger ships, and R. & F. Shipbuilding yard having now extended its building berth, it could now easily accommodate ships 400+ feet in length, so that the size of the ship being built was restricted only by the amount of land available to build on.

Subsequently, several great sailing barques were built, beginning with one – of four in total – for the shipping company of Crane Colville & Co., called *Highland Chief*. The yard was now getting into the building of much larger vessels, of more than 1,000 grt.

The majority of these large ships were being built as cargo ships for the booming imperial markets. These ships were the lifelines of the island nations of Great Britain: with routes over all the seas of the world, and under the protection of the mighty Royal Navy, which had built naval stations globally, the ships could pretty much roam worldwide secure and free.

Ships were getting bigger and bigger – and the ship owners' profits were also growing exponentially, as one ship could now do what it had taken two or even three ships previously. While the shipyards would constantly reduce the men's wages when orders were short, they would never reduce the payout to their shareholders. Within this context, R. & F. managed to break into a particularly exclusive market at the beginning of the 1880s. In some fine marketing of its product, the company persuaded the well-known owners of the first two steam yachts built to sail them around the British Isles, showing off the skills and workmanship of the new shipyard at Leith. This market was to epitomise all that was great about the shipbuilding prowess of the men from Leith in building some of the finest vessels afloat.

The market that R. & F. managed to break into was the building of luxury ships. Named, somewhat obtusely, ⏁steam yachts⏁ these vessels were like no yachts from a previous or future time; they were top-of-the-range luxury floating palaces. They were ships designed by some of the best-known names in the industry, with a line that exuded elegance and a form that in my opinion has yet to be equalled to this day

*The beautiful Auxiliary Steam Yacht* La Belle Sauvage *was built by Ramage &*
*Ferguson Shipbuilders and launched in 1894. (Author's collection)*

The fact that the Leith shipbuilders took to this new type of ship, and the building and outfitting of them, is no more of a surprise than that they did it working in terrible conditions, in all weathers, and for consistently low pay. The fundamental inequality – and irony – is that the owners of these new luxury ships were some of the wealthiest people in the new industrial age, rich people who had made their fortunes on the backs of such humbled, skilled workers, who were expected to be grateful having a job building these luxury new playthings.

Fortunately for the shipbuilders of Leith there was a large demand for such vessels as the world's wealthy tried to outdo each other in their fabulously expensive pissing competition. This meant much welcome work for the R. & F. yard, although surprisingly the company never made large profits on these ships. The work was, however, great publicity for just how good the shipbuilding in Leith was, and for generating more business. Many owners of these fabulous steam yachts also owned their own shipping lines – and might well be encouraged to place further orders for general cargo-type ships. Through this arrangement, a lot of orders were indeed won for the building of more practical ships that generated a higher profit margin.

The firm of R. & F. Shipbuilders and Engineers, later known as Ramage & Ferguson Shipbuilders and Engineers Ltd, was eventually to build upwards of 90 of these magnificent steam yachts. What is difficult to imagine nowadays is that amongst all the abject poverty and destitution that was around in Leith at the time, the craftsmen were building some of the most expensive and luxurious vessels ever to take to the sea. Not that the shipwrights or

*Auxiliary steam engine with all brass and copper polished and gleaming. (Author's collection)*

riveters would have given a shit for the owners; they were mostly interested in keeping their jobs and if possible getting some overtime to supplement their small wage. The best way to ensure this happened was to produce the finest product possible.

# FOUR: The Golden Years

During the year of 1881, the firm of R. & F. Shipbuilders built a total of ten ships. From this total came a very interesting order from the well-known shipping line of Walker, Donald & Co., which ordered two large single-screw steamers from the yard. Chapter 3 included a description of the ill-fated *Craigrownie*, designed and built specially for the Spanish fruit and mineral markets, but her slightly larger sister ship, known as *Craigallion*, Yard No 29, launched on 17 May 1881, was to have a somewhat more long-lived and adventurous life at sea. In 1884 she was abandoned as a total loss during a storm in the Bahamas before being salved by an American salvage crew and taken as a prize. Refitted and seaworthy, she was renamed SS *Ozama* and was to continue her life at sea under her new American owners. Ultimately, the wreck of a ship found in 1979 was officially identified as SS *Ozama*, ex-SS *Craigallion*, by Dr Lee Spence, a well-respected wreck-finder, off the coast of South Carolina.

Two weeks after *Craigallion*'s launch came yet another fine iron steam yacht, *Iolanthe*, Yard No 24, to be renamed *Amalthea*, then *Iolaire*. Her end was one of the most tragic events of the First World War era. She was sailing from Kyle of Lochalsh to Stornoway on New Year's Eve 1918, and with the war now over she was bringing home about 300 lads who'd survived the horrors of the trenches, all looking forward to seeing their families waiting to greet them – only to be driven by a Force 8 gale onto the Beasts of Holm, rocks at the very entrance to Stornoway harbour. The trauma to the families of the 200 drowned (a significant proportion of the population of the isle of Lewis) was such that it has taken 100 years for the event to be acknowledged openly.

Yard No 28 then being launched on 30 April 1881, it was apparent that the ships were being built in a short time. The ferry passenger ship *Midlothian*, Yard No 30 – not to be confused with the earlier ship of the same name built by S. & H. Morton in 1871 – was a paddle steamer at around 920 grt, ordered by the North British Railway Company. She was launched in August 1881. Perhaps due to the opening of the Forth Railway Bridge, she became surplus to requirements and was sold on to foreign interests, who turned her into a barge in 1890. Yard No 31 came soon after, another well-known vessel, SS *Ardangorm*, which under her new name featured on a postage stamp issued by the island of Mauritius.

*Secunder, ex-Ardangorm, Yard No 31, from the R. & F. Shipyards at Leith, in this postage stamp issued by the Mauritius government to celebrate the service she rendered to the island.*

The *Candace*, at 268 gross tonnage (Thames measurement) was the sixth steam yacht that R. & F. built in 1881. She was purchased by the Admiralty and promptly renamed *Fire Queen*, being used as a tender for HMS *Duke of Wellington*. She was a strange ship, to say the least – or not so much strange as hopelessly obsolete, being originally designed as a 131-gun wooden sailing ship. She was converted to steam and was then only used as a ceremonial ship until she was broken up in 1904.

Around the year of 1881, the famous Clan Line shipping company had decided on expansion. R. & F. Shipbuilders were allocated two of the new ships, large steamers at 2,956 grt and more than 330 feet in length. This was a big order for the Leith yard to win, especially since other ships ordered by the Clan Line were being built by rival companies in the line's home town of Glasgow.

The following year, 1882, the yard of R. & F. was to build one ship at over 3,000 grt, two more at over 2,000 grt, and another two vessels at more than 1,000 grt. R. & F. also had three other smaller vessels on the stocks. The *Clan MacGregor*, Yard No 36, at 3,007 grt was the largest of these ships. She was only 20 tons gross more than her sister ship *Clan MacKenzie*, Yard No 35, which was built at the same time. They were launched in August and April 1882 respectively. The *Craigton*, Yard No 37, at 2,007 grt, was the third, and she was launched on 29 September 1882. She was an order from the Glasgow Steamship Co. Ltd, with Walker, Donald & Co. as managers of the ship. Her principal dimensions were a length of 275 feet with a beam of 37 feet and depth moulded of 23 feet 6 inches. With engines supplied by Blaikie Brothers of Aberdeen, her three-cylinder triple-expansion engine would produce a nominal 200 hp. She was to end up as an Italian ship, renamed SS *Fratelli Prinzi*, when sold to Prinzi Bros in 1900, and would eventually run aground and be wrecked in July 1908.

The two-deck steamer *Clan MacGregor* had a relatively short working life of only seven years before she collided with the Danish steamer SS *Cathay* and sank off Cape St Vincent in September 1899, on a voyage from Madras to London. By an extraordinary coincidence – if there is such a thing at sea – both ships had been built at Leith by the same shipbuilder. SS *Cathay* had been built at the R. & F. Shipyard as Yard No 156 and launched in May 1898; she was to be lost sometime later, during the First World War.

These large ships were all completed on time and within budget. Thus the relatively new shipyard of R. & F. was gaining a reputation for fine craftsmanship, and for producing ships with a high standard of workmanship – as good if not better than the longer established S. & H. Morton shipyard next door. This confirmed R. & F. as a serious player in this very competitive industry. These were good times for employment at the yard, as repair and salvage work were also in demand. (See also the previous story about the steamship *Britannia*.) From its incorporation as a limited liability company in 1892, the firm went on to build deep-sea freighters and coasters, and continued with the magnificent steam yachts for which it was to gain worldwide renown.

In fact, this whole last quarter of the 19th century – steam overtaking sail power and steel replacing iron – was one of prosperity for shipbuilding in the British Isles as a whole, with record tonnage outputs from both the new and the established shipyards.

The years of 1883 and 1884 were the years of the large sailing ships down Leith way, and R. & F., the firm that had begun the previous year with the building of the aforementioned *Highland Chief*, Yard No 32, would go on to build four more of these magnificent vessels, two of them designated as barques and two of them as full-rigged sailing ships. These were followed by the barque *Highland Glen*, Yard No 41, of some 1,032 grt, built for the same owners, Crane, Colville & Co., and launched at Leith on 23 November 1882.

With the steel-built *Loch Ard* then coming in as Yard No 39, and launched on 23 December 1882, R. & F. moved firmly from building iron ships to steel ones. As such, the *Loch Ard* was a very significant ship in the history of the R. & F. Shipyard. Sadly, renamed the French vessel *Tunisie*, she would find herself wrecked on Lundy Island in the Bristol Channel, and finally broken up for scrap at Cardiff in February 1892.

During the year of 1882, the firm of Ramage & Ferguson launched a total of nine ships of which four were screw steamers, four were steam yachts and one a sailing barque, for a total aggregate in tonnage of some 12,100 tons. Then the next year was to be another stellar one for shipbuilding all around the British Isles. R. & F. Shipbuilders would again have a reasonable share of this shipbuilding bonanza, beginning with the launch of the large iron three-masted, full-rigged sailing ship of some 1,686 tons, *Mount Carmel*, Yard No 43. She was lost in July 1916 when she disappeared during a storm off the Florida coast.

The iron steamer *Ecossaise*, Yard No 44, was launched in March 1883, a standard steamship of 849 tons, and one that was to have a long working life – 77 years. She survived the two world wars and was only broken up for scrap in 1950, having had four different owners during her time. She also went through a few name changes and carried SS *Caring* (given in 1922) when she went for scrapping at Charlestown, New South Wales. Following on, the iron steamer *Granada*, Yard No 45, of 958 grt, launched the month after her, did not last anything like as long as the *Ecossaise*, ending up wrecked on 'Normande Rock' (probably off Normandy) in March 1888.

As the year continued, a very large full-rigged ship of some 2,079 grt was launched in July: *Earl of Shaftesbury*, Yard No 46. She had four masts and a large cargo-carrying capacity, which would be utilised for the jute trade to India. She was wrecked off the coast of Ceylon in May 1893, almost ten years to the month after she had been launched. There are some gory tales about the crew, and the locals' lack of attempts at rescue as the ship lay helpless on the rocks.

The next notable launch of an iron steamer was that of the 1,082-ton SS *Carriedo*, Yard No 47, which was built for Reyes & Co. of Manila and launched in August 1883. She had an interesting history spanning some 40 years, and as USS *Manila* even became part of the U.S. Navy:

SS *Carriedo* was a passenger/cargo steamer which was destined to be a mail ship for the Spanish mail service between Singapore and Manila. With her high level of passenger accommodation, she was the smartest and also the fastest steamer owned in the Philippine Islands. The Spanish government purchased her in 1886 and assigned her as a transport to their naval forces based in the Philippines.

As the Philippines at the time were under Spanish rule she was caught up in the Spanish–American war which although only lasting for some three and a half months was to have a profound effect on the outlooks of Spain and America. The obsolete Spanish fleet of which the *Carriedo*, by then named *Manila*, was part of was sunk in Manila harbour, and with the cessation of hostilities the Americans discovered that the ship was only partially sunk. The ship lay in 3 feet of mud near the navy yard, and her officers begged that she not be destroyed because she was unarmed and was a coast survey vessel.

So she was spared, and on 4 May 1898 the Americans towed her off, raised steam on her, and anchored her near the squadron. She proved to be a handsome steamer of about 2,000 tons and was laden with supplies and 500 tons of coal, from which the cruiser *Raleigh* supplied her bunkers. Taken into the U.S. Navy, she was first commissioned on 20 July 1998 and attached to the Asiatic Station. During the Philippine Insurrection, she was actively engaged in patrol duty, convoying troops, cruising on station etc.

She participated in the bombardment of San Fabian on 7–9 Nov 1899 and sent a landing party ashore on 16 Nov 1899 during the taking of Zamboanga. She was station ship at Cavite during the second quarter of 1901 and the first quarter of 1902, and was at Mare Island on 30 Jun 1902. She was placed out of commission at Mare Island on 1 July 1903.

*Manila* was described in Navy annual reports as a transport in 1899 to 1903 and as a gunboat while in reserve in 1904–1906.

She was converted to a prison ship at Mare Island in 1905–1906 and was fitted with 38 cells. Stricken and ordered sold in 1913, she was offered for sale by the Navy Department on 27 Jan 14 with bids to be opened 4 May 14. She was sold to J.W. Strong and re-entered merchant service in the Far East.

*(Navsource.org)*

Renamed SS *Wanli*, in 1920, the former *Carriedo* was on a voyage from Dalian to Shanghai with beans and cereals, when she was sunk in a collision on 21 May 1923, at Tsin Shan Tei, on the Shantung peninsula. There was the loss of 16 lives, including passengers and crew.

The steam yacht *Merrie England* followed next as Yard No 48, launched in June 1883, followed by SS *Craigendoran*, Yard No 49, another iron cargo steamer of 1,471 grt launched in October of the same year. An order from Walker, Donald & Co., she was with that company for 15 years before being sold on to foreign interests, where she was renamed twice; finally,

*The ex-SS* Carriedo, *renamed USS* Manila. *(Navsource.org)*

under the name of SS *Espoir*, she was involved in a collision and sank off the coast of Turkey in August 1924.

SS *Crosshill*, Yard No 50, was another noteworthy Leith-built ship, and one which became part of a well-known maritime court case. She had been launched in October 1883 for the firm of McBeth & Grey, which owned her up until 1900, when she was sold on to foreign interests; she was sold on once more, in 1913, to A. Fahrenheim, when she became SS *Horst Martini*. So as she was a German ship at the outbreak of war, she was detained in Wales and subsequently turned over as a war prize to the British government, which ran her as a collier, very much in aid of the war effort. Then in May 1915, involved in a collision in fog with SS *Runic*, a huge White Star liner of around 12,000 grt, she was clearly going to come off worst, and she sank just off Beachy Head. This presented an interesting case (great for the lawyers of the day who could increase their daily fees while everyone tried to sort out the proceedings) when the owner, master and crew of the *Horst Martini* – the owner being, in fact, the government – brought a case against White Star. The laws of the sea can be even more obtuse than the laws on the land, and at the end of the day the *Horst Martini* was adjudged to be three-quarters to blame for the collision.

R. & F. Shipbuilders built seven vessels during 1884 with a lower tonnage than the previous year. The largest, at 1,040 tons, was the three-masted steel barque *Highland Forest*, Yard No 51. The launch of *Highland Forest*, mentioned by the famous maritime author Joseph Conrad in his book *The Mirror of the Sea*, was followed by the build of two steam yachts and the paddle steamer *Bournemouth*, Yard No 54, which was wrecked on the rock of Portland Bill in 1886.

Next on the stocks was another steam yacht which was to become quite famous. Named the *Lady Aline*, Yard No 55, she was to become a grand old lady of the seas, with a life well into her seventies when in June 1960 she was lost, reported as missing and never seen again. In her time, she had been HMS *Oberon*, then HMS *Hawke* when used by the Admiralty from 1888 to 1906, before being sold out of service under the name *Undine*. Then used in a fisheries and customs role, she was even occasionally used as a royal yacht for Queen Victoria.

Then followed the steam cargo vessel *Bordeaux*, Yard No 56, 550 grt, and another vessel that was lost as she foundered near the mouth of the Humber on a voyage to Methil in Fife during the winter of 1897. Then the final vessel of 1884 was the twin-screw tug *Otter*, Yard No 57.

# 1885 – 1889

By the mid-1880s even the British Admiralty had to concede that steel was the ideal material to be manufacturing future warships – although it appears that the good Lords of the Admiralty only became convinced of this due to the grounding of one of its new warships. To their absolute astonishment, she never sank. After inspection was carried out in dry dock, this was found to be because she had been built from steel; only some of her bottom plates had buckled, with no loose rivets to be found.

Siemens, with a couple of other perhaps less well-known processors, was now producing steel with ductility, toughness and strength, that could be bent cold and formed into the required shapes, all of which made it much more suitable for shipbuilding. Also, it would soon be cheap enough, and of sufficient quality and size, to take over completely from iron, with galvanised mild steel as the material of the future.

SS *Mascotte*, Yard No 58, was a large steel steam cargo ship, powered by compound surface condensing engines. With a large carrying capacity, at some 1,100 grt, she had a length of 250 feet with a beam of 30 feet 6 inches and 16 feet 2 inches in depth, with a single screw powering her to a speed of 13 knots at her sea trials – above the required contract speed. She was launched on 15 January 1885, built for the Gibson Line of Leith for trade between Leith and Rotterdam. Unfortunately, she was another casualty of the First World War, with the loss of one of her crew on 3 September 1916, when she hit a German mine which had been laid by UB-6 off the coast of East Anglia.

SS *Malacca*, Yard No 59, was an iron screw steamer designed for passenger and cargo use, built and engined by the firm for Kim, Seng & Co., Singapore. With a length of 180 feet and beam of 26 feet 6 inches and a depth of 16 feet, she was 653 tons, and her engines produced a nominal horsepower of around 125. Launched in January 1885 she went on sea trials on 7 March, easily exceeding her service speed with around 11½ knots over the measured mile in the Firth of Forth.

April 1885 saw a very nice steam yacht launched, which at one time would be owned by the Imperial Russian Navy. At her launch she was named *Katrena*, Yard No 60.

The next ship featured shows just how quickly the shipbuilders were building ships at that time. Another iron screw steamer, she was also another order from the firm of Walker, Donald & Co., Glasgow – the fifth such ship to be ordered by the same company. To be named SS *El Calloa*, Yard No 61, she was a fair size at 1,000 tons, 210 feet in length with a beam of 31 feet. She had a moulded depth of 36 feet to her upper deck, supplied with the builder's own engines and specially designed for the South American fruit trade. She had a light draught to enable her to travel up the Orinoco river at all times, and was fitted with accommodation for 20 passengers. She also carried all the latest equipment as required by the American Board of Trade, including fire extinguishing apparatus in every compartment in the ship. The *El Calloa* achieved a speed of 11¼ knots during her sea trials, which was in excess of her contract speed, and from the signing of her build contract to her handover was a period of only five months.

The next to be launched was the stern-wheel paddle steamer, *Henry Vinn*, Yard No 63, built to an order from the Church Missionary Society, London, for service on the River Niger. She was to be shipped out to Africa on board one of the African mail steamers from the port of Liverpool. This launch was quickly followed by the small steam launch Yard No 64, *El Strivido*.

The barque *Crown of India* was launched as Yard No 62 on 30 June 1885 for an order from Robertson, Cruickshank & Co., Liverpool. She was four-masted and built of iron, and, at some 1,970 tons, a large ship. The full-rigged iron ship *Crown of Italy* was then launched as Yard No 66 on 24 September 1885 for the same Liverpool owners. As soon as she was fitted out, she was off to Frederickstadt (now Friedrichstadt) to take on a load of timber for cargo to voyage to Melbourne, Australia.

Ramage & Ferguson had done relatively well – again – during 1885, launching a total of nine vessels, two of them sailing ships of a large tonnage. Out of those nine ships, two were constructed from steel while the others were iron ships. Only one of those described –SS *Mascotte* – was built for a foreign owner; the rest were all for British owners. By contrast, the firm of S. & H. Morton & Co. had built only one, the *Britannia*, for the British.

It was not just in the building of ships that the firm of Ramage & Ferguson was heavily involved, as in common with John Cran and other engineering companies around Leith, it also built marine engines and boilers. This gave it the added advantage of being able to furnish the ships that it built with its own engines – unless the order specified that some other engine was required. Out of the nine ships built by R. & F. in 1885, the only two that did not have engines built by the firm were the two large sailing ships.

The following year was to begin with the launch of the large three-masted steel barque *Highland Home*, another ship for Crane, Colville & Co. of Glasgow. She was the fourth large sailing vessel ordered by the same company, and she was launched on 16 February 1886. The yard also built and launched a couple of steam-powered ferries for work in the Far East and Australia, the first of them being the steamer *Beaver*, Yard No 71, launched in 1886.

*The steamer* Beaver *under way. (State Library of Queensland, Australia's John Oxley Library)*

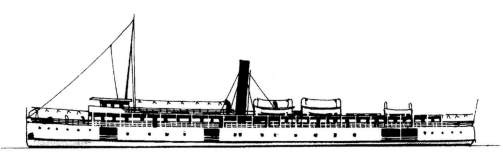

Fatshan *(Author's collection)*

The yard built and launched a shallow-draught ferry for China named *Fatshan*, Yard No 76. Built and launched in 1887, she was one of the first ever ships to be designed with a twin screw, revolutionary at the time as an entirely new concept – and one which was to be quickly taken up by many other builders. This concept was also a real challenge to her designers and to the loftsmen who would have to develop her shape; the mind-bending complexity of developing hull plates through not only a double lateral curvature but also to fit a curvature in the fore and aft direction would give many a loftsman of the time a headache.

The *Fatshan* had a triple-expansion engine, which again was a very new concept; the first of them had only been introduced some six years earlier. She was in fact way ahead of her time, evident even today when looking at a photograph or a drawing of her.

*Fatshan* had not only first-, second- and third-class cabins but also steerage accommodation and un-berthed passenger space. She was the latest design in a steamer, built at a cost of £22,000 – Ramage & Ferguson having undercut Scott & Sons of Bowling's quote by £4,000. She was one of the very first vessels to be so very well equipped throughout. In today's climate of safety-consciousness, it is perhaps interesting to note that *Fatshan's* simple steering gear, which consisted of a series of rods connecting the wheelhouse with the aft quadrant above the rudder shaft, was not completely boxed in over the engine casing. Those moving rods were, in fact, completely exposed at ankle height along the port and starboard sides of the upper deck – a definite hazard to the farepaying public, since the vessel could carry up to 1,000 deck passengers. In December 1933, *Fatshan* was sold to breakers, having been replaced after 45 years of service by her namesake, *Fatshan II*.

To complete the builds for the year 1887, three steam yachts were manufactured and launched after *Fatshan*, along with a yawl named *Atlantis*, and then a small steamer called *Chamroen*.

Yard No 82 was to be the steel screw steamer *Hailoong*. She was built to an order from the Douglas Steamshipping Company, Hong Kong, which required a cargo/passenger ship for use in the Far East. This modern fast ship, with a gross tonnage of around 1,253, was launched at Leith on 29 February 1888, and was intended for use between Hong Kong and Formosa (now Taiwan). She was launched lightship without her masts and engine, which would be fitted at the Albert Dock. She was 230 feet in length, with a beam of 33 feet and a depth moulded of 21 feet, and she was to be fitted with triple-expansion engines, supplied by the builders, to produce around 1,300 hp. She was certified to the highest class at Lloyd's and passed by the Board of Trade to receive her foreign sea-going certificate.

With her decks and other woodwork of teak, she was a very high-class passenger ship, also capable of carrying a fair amount of cargo, as her trade route demanded. On her sea trials carried out in the Firth of Forth she passed all the required tests and attained her required speed; on the measured mile she exceeded her service speed, to sustain 12¾ knots. Soon after her trials she would be on her way to Hong Kong to begin her service life, and she was used on this route for some 17 years before being sold and renamed *Schleswig*, still working out of Hong Kong. She was sold on a few more times, eventually ending up with a Japanese shipping company and renamed *Ichi Maru*. When surplus to requirements, she was broken up in Japan in 1931 after some 43 years of a useful working life at sea.

The next two months at the shipyard would see the launch of another two fine steam yachts, the first to be launched in March 1888 as *Red Eagle*, Yard No 83, a yacht designed by the well-known luxury yacht designer, St Clare John Byrne. She was followed down the slipway in April by the steam yacht *Garland*, Yard No 84. However, whilst the build of the luxury steam yachts was great for showing off the prowess and skills of the shipbuilders, it was the more mundane steamers that were the bread and butter of the shipyard.

Next on the stocks was the 1,245-ton SS *Ancona*, Yard No 85, 249 feet in length by 33 feet beam with a moulded depth of 18 feet. She was launched in June 1888, having been built to an order from the well-known local Leith shipping company James Currie & Co., a firm that was to go on to be a very good customer of the relatively new shipyard. In keeping with the look of ships of the time, and despite the large increase in the use of steam power, she still had sails; she was schooner-rigged with three masts. She was to be used in the company's continental trade. Powered by a triple-expansion engine fitted by the builders, she was completed and handed over to her owners the following month.

SS *Ancona* was sold on from James Currie in 1911 to J. Hall & Co., London, but she did not survive the First World War, being sunk by a German submarine on 28 May 1917. She went down some 110 nautical miles south-west of Ushant with the loss of all 25 of her crew plus a passenger. SS *Ancona* had been the lead ship in a two-ship order from James Currie, the second ship to be launched on 9 August 1888. SS *Ravenna*, as she was named, stayed with Curries for 13 years before being sold on, in 1911, to the London & Edinburgh Shipping Co., Leith. She was involved in a collision with the Swedish steamer *Ulla* in August 1917 with the unfortunate loss of five of her crew.

The next steel steamer built was a good example of how things were starting to change in ship design. She was SS *Maristow*, Yard No 87, and at 1,679 tons she was a fair-size ship with a length of 263 feet. Designed to operate primarily in the Mediterranean and the Black Sea, she was an order for Bellamy & Co. of Plymouth, with a long bridge deck extending forward to her foremast and with a cellular double bottom construction forward and aft for water ballast. SS *Maristow* was launched from the yard on 1 December 1888, powered by triple-expansion engines built by the yard, and producing a nominal horsepower of 160. She easily achieved more than her contract speed during sea trials, being timed over the measured mile in the Firth of Forth at 11¼ knots.

The launch of SS *Maristow* strengthened the positive outlook for the following year. The yard was also close to completing a large steel sailing ship, and the keel had been laid for another steamship on order as well as working on the build of the largest steamship to be built at Leith at the time. With day and night shifts being worked at all the shops of the Leith shipyards, and repair and conversion work in hand, 1889 looked like being a good year. Business was looking good, so R. & F. went ahead and expanded the boiler shed to double its size.

SS *Suffolk* was built as Yard No 88 in March 1889; she was a steam cargo ship of no less than 3,303 grt with a length of 330 feet, the largest ship built at Leith up to this time. She was powered by triple-expansion engines built by the shipbuilder, and complete with all the modern conveniences at the time for the handling of cargo and the comfort of crew and passengers; her owner, Money Wigram & Sons Ltd, was one of the oldest and hence most experienced firms on the London to Australia routes. During her sea trials she easily surpassed her guaranteed 10-knot contract speed, and she was then handed over to her new owner.

After this large steel steamship had been used on the London to Australia route for a couple of years, she was sold on to a Spanish company in 1893/94 and renamed *Berenguer el Grande*. She continued working for that owner until, named *Giacomo*, she was broken up in 1911 in Genoa, Italy. The later shipyard of Henry Robb Ltd would have been very happy to be building ships of this size some 60 or 70 years on.

SS *Suffolk* was swiftly followed by the launch in April of another steel steamer, SS *Rosary*, Yard No 89. Then in May came the luxury steam yacht *Semiramis*, Yard No 93, designed by A.H. Brown, the well-known naval architect and designer. The rest of 1889 was taken up with the building of a further five steamers, making for a busy time in the shipyard. Two of the ships on order were for the firm of R. Conaway & Co. of Liverpool, to be named SS *Realm* and SS *Rex*, both of approximately 1,750 tons and 265 feet in length.

Both ships were known as deck steamers, due to the new designs of such steam cargo vessels. Another interesting thing about SS *Realm* was that although she was still rigged as a topsail schooner, her masts were perpendicular rather than raked, enabling her cargo derricks to work much more efficiently than those set up to a raked mast.

With the build and launch of SS *Barraclough*, Yard No 92, as well, the yard had many demands on labour: four ships in the construction stage at the same time, plus repair contracts to fulfil. The shipyard next door at S. & H. Morton had two ships under construction, so it was a busy and noisy foreshore at the start of the Trade Holidays in the first two weeks of July. Meanwhile, J. Cran & Co. was being kept busy on a large repair job on SS *North Star*, one of the James Currie fleet of ships,

This work at the R. & F. shipyard was to be followed by a two-ship order from James Currie. The first of the two was SS *Zamora*, Yard No 94, a cargo steamer of 249 feet in length, and she was launched on 25 October 1889. *Zamora* would only last until 1897, when she was wrecked in the December of that year in the Nieuwe Diep, the lake by Amsterdam.

The second of the ships, SS *Weimar*, was slightly larger, as she was designed as a passenger ship for use on the ship owner's Leith–Hamburg route. She had a length of 254 feet with a beam of 34 feet and a moulded depth of 25 feet, with accommodation for 56 passengers in first class, 16 in second class and 170 in steerage. Her triple-expansion engines were supplied by the builders and showed power of around 1,350 hp. She was launched on 21 December 1889, and during her sea trials she attained a speed over the measured mile of 13¼ knots, more than any other of the Currie ships. With a very high standard of workmanship, she was indeed the Queen of the Fleet, and she would stay with the same owners. Surviving the First World War, she continued working until broken up in 1933.

The launch of SS *Weimar* concluded a very busy year for the shipyard, with good prospects for the new decade. Not since 1883 had there been so much work for the shipbuilders of Leith. The industry was well accustomed to the pendulum of good and bad, but for now all was good.

*SS Weimar, Yard No 95 (University of Glasgow Archives & Special Collections, Currie Line collection, GB 248 UGD 255/5/7/35)*

# 1890-1900

With launches and sea trials taking place from the first month, and new orders being placed for everything from large sailing ships to small steel steamers and luxury steam yachts, plus engine orders and ship repairs, the new decade began with renewed optimism at the shipyards of Leith. And it was an optimism shared around the country, day and night shifts being worked as and when required. R. & F. Shipbuilders had a total of eight ships on its order book for the coming year, and S. & H. Morton had a total of three ships on order.

February was the month that saw the twin-screw passenger steamer *Heung-Shan*, Yard No 96, go down the ways; she was another order for the Hong Kong, Canton & Macao Steamboat Co. In many ways very similar to her predecessor, the very popular *Fatshan*, she was much larger and faster at 1,985 tons and she attained a speed of more than 15 knots on her sea trials. The *Heung-Shan* was designed to carry a large number of passengers, complete with opium as a cargo, and was outfitted to a very high standard with two triple-expansion engines adapted for high speed on a shallow draught. At some 300 feet in length, with a beam of 54 feet and a depth of 28 feet 6 inches to her upper deck, she was a fine vessel.

While it has to be said that the shipbuilders of Ramage & Ferguson constructed some of the finest ships to ever sail the oceans of the world, some of the ships built at the old Leith yard had their fair share of tragedy attached to them as well, and while the maritime world is no stranger to tragedy, the tale of this next ship built at the Leith shipyards of Ramage & Ferguson is a very sad tale indeed.

The captain and owners of SS *Talune*, Yard No 97, along with a resident islands doctor, were responsible for more deaths than occurred in some of the battles of the First World War. She was launched into the cold waters of the Firth of Forth in April 1890, and at 2,087 tons with an overall length of around 230 feet, she was a fair size of steamship for her day. She was an order for the Tasmanian Steam Navigation Company of Hobart, coal-fired, and with a triple-expansion engine turning her single screw. With accommodation for up to 175 passengers and a crew of 56, she was initially intended for service on the Hobart–Sydney run. (The Tasmanian Steam Navigation Company was taken over in 1890 by the Union Steamship Co. of New Zealand, a company which was to be a very good customer of the Leith shipyards over the years).

SS *Talune's* accommodation and the workmanship on her interior were of a very high standard, and she had all the latest modern appliances: a very well fitted-out ship. She worked the passage between New Zealand and Australia before going onto the run between New Zealand and the Pacific Islands. Her early years of voyaging were pretty run-of-the-mill for such a ship, although she was involved in two salvage operations; one being the tow of a ship twice her size which had been adrift and missing for eight weeks: the *Talune* towed SS *Perthshire* into Sydney to much fanfare and good wishes. She also saw service during the First World War when she was taken over by the New Zealand authorities to be used as a military transport ship and given the nomenclature of HMNZT 16. She was used to transport at least one contingent of troops to Western Samoa, and at some point in 1916 she was relinquished from her military duties and reverted to civilian use to resume her regular voyages between the Pacific Islands and New Zealand.

Shortly after that, in March 1917, she suffered her only stranding when she struck uncharted rocks in the Egeria Channel, Tonga, while carrying general cargo and 15 passengers. While a fair bit of damage was done to her plates and frames, the official Court of Inquiry found her master, J. Morrison, clear of any wrongdoing. On 7 November 1918, shortly before the declaration to end the war in Europe, SS *Talune* arrived at Apia in Western Samoa on one of her regular voyages from Auckland (she also called at ports in Fiji, Tonga and Nauru and then back to Fiji again before returning to Auckland). At that time, the islands of Western Samoa were administered by New Zealand, having been seized at the start of the First World War. Along with the U.S. administering the eastern islands of Samoa, this part of the world was clear of German forces – but as it happened the Germans were to be the least of the problems for the people of these Pacific islands.

At the time of the ship's departure from the city of Auckland, pandemic influenza – Spanish Flu, as it's widely known – was spreading rapidly in New Zealand, resulting in many deaths. Before leaving, two crewmen had reported sick and were sent ashore, but by the time *Talune* reached Suva in Fiji on 4 November, several more crewmen had been infected. Although the ship remained officially in quarantine alongside the wharf, as none of the Fijian passengers were stricken they were allowed ashore and the cargo was unloaded. Not only that, but as was the custom of the time, about 90 Fijian labourers were taken on board to work the cargo as the ship proceeded on its planned voyage. By the time the *Talune* reached Apia on 7 November, most of those Fijian labourers were ill. The ship's quarantine at Suva was apparently not mentioned on arrival in Apia, and the acting port health officer there was not aware of the epidemic in Auckland. After what seems to have been a somewhat cursory examination the ship was granted pratique.

With the passengers allowed to disembark, *Talune's* captain told the medical officer, Doctor Atkinson, that there was nothing serious on board. However, by 31 December, at least 7,542 Samoan people had died from virulent influenza, and further deaths from influenza continued into 1919.

A commission of enquiry calculated a final death toll of about 8,500 – about 22 per cent of the population of Western Samoa. While the impact of the pandemic was undoubtedly amplified by the Samoan cultural response to illness, which required the family to gather around a sick person, the New Zealand administrative response to the pandemic was at the very least inept.

SS *Talune* went on from Apia to Tonga (calling at Neiafu, Vava'u, Ha Apai) and to Nuka' Alofa in Tongatapu, where she arrived on 12 November 1918. Within a few days of the *Talune's* arrival, the disease had spread, with heavy loss of life; estimates vary that between 1,800 and 2,000 died. After Tongatapu, she sailed for Naura, where once again the first cases of influenza appeared ashore within a few days of her departure. Few, if any voyages in history, whether in peace or war, have resulted in so many deaths in such a short time. The impact of the *Talune's*

*The 'death ship', SS* Talune, *Yard No 97. (Author's collection)*

deadly voyage is still remembered today, and has influenced pandemic disease planning into the 21st century.

There is no record of what happened to the ship after her disastrous voyage until 1921, when the Union Steamship Company records show she was laid up. Then in 1925 she was hulked and in November of that year she was filled with rocks and scuttled to form the foundation of a breakwater at the New Zealand port of Waikokopu, now no longer in use as the wharf and the breakwater have been reduced to rubble by the constant southerly swells. It's a very sad end to a sad tale, a disaster that was wholly the responsibility of the ship's master and the medical officer, who between them agreed to bring down the yellow flag and give the unfortunate ship a clean bill of health.

Meanwhile, back home in the Leith shipyards work had been ongoing; steamships of iron and steel were in production. But also the mighty windjammers racing each other around the fearsome Cape Horn included some of the ships built here. The *Drumrock* and *Procyon* were two of the largest three-masted barques ever built. The *Trade Winds*, the *Crown* ships, and the *Highland* ships for Crane, Colville & Co. were amongst many other examples. Some of the very fine sailing barques built here were used in the booming trade routes from Chile to Europe. Their cargo comprised valuable nitrate, something that was needed for use as fertiliser and, of course, the production of ammunition – which helps us to understand why some of the famous German shipping lines were so active in this trade at that time. In 1883, the four-masted full-rigged *Earl of Shaftesbury* was built and launched. She was a large ship of 2,079 grt, just over 280 feet in length, and she was to plough a fine furrow through the high seas for ten years before she capsized and sank on a voyage from Bombay to Diamond Island, foundering off the south-west coast of Ceylon (now Sri Lanka).

It is clear, then, from these and other examples in earlier chapters, that even with the huge strides in development and use of steel, there was still a market for some of the long-haul trade routes to use huge ships powered by the wind. And, it's equally clear that whatever the material of production, no ship was ever more powerful than mighty Neptune, especially when his wrath was aroused. Many of these ships did have long and productive lives at sea, but one thing noticeable about these – and earlier – times is the number of vessels lost, frequently with the loss of the entire crew. Bad weather, running aground, being wrecked on some rocky and poorly charted coast, in collision … There's no doubt that going to sea was, and still is, one of the most dangerous professions. Although much *has* changed in the last couple of hundred years, the sea is still as unpredictable as it has ever been; woe betide anyone who does not give the water the respect it so richly deserves.

Shipbuilding itself could not be taken for granted, either. It was always cyclical: an industry in which the workers could, by working faster, work themselves onto the street, an ironic point never lost on shipbuilders. With the build of a ship only taking around three months, and with four berths to keep building on, the firm of R. & F., for example, while appearing to be busy, required a large back order of vessels on the books at any given time in order to plan ahead for materials and labour. It also had the engine works to keep going, and the repair orders for ships.

In July 1890, following on from the annual Trades Holidays, the industry was facing a general slowdown on the Clyde, although work *was* still progressing at Leith, and there were some more very fine steam yachts built in the shipyards in this decade; wealthy clients such as the King of Siam were still putting forward orders. Most of their vessels designed by one of three well-known designers of the day: Cox & King; G. L. Watson (still in existence today in Liverpool, having begun in Glasgow) and the naval architect St. Clare John Byrne (1830–1915).

Inevitably there were ongoing rivalries. The steam yacht *St George* was quietly launched as Yard No 100 on 30 August 1890, but of more interest locally was SS *Nevada*, Yard No 101, launched from the R. & F. yard on 13 August 1890, and her sister ship, SS *Morena*, Yard No 60, launched from the S. & H. Morton yard on the same day. It wasn't unusual for ships to be built in different yards – but these two shipyards were immediate neighbours. With both ships on order from Currie, there must have been a hugely competitive element – but was it by accident or design that the two ships were launched with only half an hour between them?

The last ship R. & F. launched that year was SS *Tyne*, launched on 15 November 1890, with trials completed on 28 January 1891. She was built for the Royal Mail Steam Packet Co. of London, and her home port was to be the island of Antigua in the West Indies.

With so much work on R. & F.'s order books, all was not well with the men when the time came for the firm to set out the winter working hours. It insisted on a 54-hour week. While this was the normal working week when the yard was busy, the firm of R. & F. Shipbuilders was insisting that on Saturday the men should work until quarter to two instead of the normal twelve noon finish. The consequent strike went on for a full two weeks before the men eventually gave in and went back to work on the company's terms.

*SS* Nevada, *Yard No 101, undated (University of Glasgow Archives &*
*Special Collections, Currie Line collection, GB 248 UGD 255/5/7/29)*

Overall, R. & F. Shipbuilders were at number 48 in the 1890 tonnage table out of 67 shipyards in the British Isles, even though that year its tonnage launched was down by approximately 1,000 tons compared with 1889. In addition, Russell & Co. on the Clyde had overtaken Harland & Wolff as the main builders in 1890.

As 1891 began, the shipyards were faced with an unwelcome extended holiday time due to an ongoing national railway strike that all but paralysed the main industries; they relied on a plentiful supply of coal and steel, as well as all the other materials required to build ships. The strike was to continue well into February, and things were slow in the three big Leith shipyards sited along the foreshore to the west of the west pier. It did not, however, prevent the launch of Yard No 102, which was entering into the water as *Ladye Mabel*, launched on 12 January. This was to be the year of the large sailing ship in Leith, beginning with the large four-masted barque *Trade Winds*, Yard No 104, at 2,859 tons, launched in February, followed the next month by SS *Moscow*, Yard No 105, for the famous shipping line later known as the Ben Line, (known locally as William Thomson & Co.). SS *Moscow* went on to be requisitioned by the Russians during the First World War, and was scuttled by the Bolsheviks at Petrograd on 21 October 1918, after the Russian Revolution. She was the last of the Leith-built ships to be lost during the First World War.

Two more steam yachts were then launched in 1891 before the large steel four-masted barque *Drumrock*, Yard No 108, would take to the water. The first of the two was an auxiliary schooner-rigged yacht for Count Stroganoff, the *Lady Zara*, designed by W.C. Storey and finally launched in July. The second, smaller, one was named *Lady Ina*, Ship No 107, ordered by John Anderson of Glasgow, designed by St Clare John Byrne, and launched in May.

The *Drumrock* at 3.182 tons would, when sold to F. Laeisz, Hamburg, in 1899 and renamed *Persimmon*, go on to become one of the famous Flying P liners of 'round the Horn' fame. She was used in the nitrate trade between Chile and Europe. Built between her construction and another large steel barque, *William Tell*, was the screw steamer SS *Corunna*, Yard No 109, yet another of a two-ship order for the Currie fleet of ships; Currie was becoming a very welcome large customer of R. & F. Shipyard. SS *Corunna* was launched in August 1891.

At almost the same time as the build of the 3,107-ton *William Tell*, another very large ship was being constructed. She was the *Procyon*, one of the largest ever full-rigged ships. Built at 2,120 tons, she was launched in December 1891, just one month after the *William Tell*.

Into 1892 now, with the launch of SS *Bernicia*, Yard No 112, in February. She was yet another fast passenger/cargo vessel ordered by Currie & Co., and was designed for the company's Leith–Newcastle route. Despite being almost yacht-like in appearance, her accommodation and arrangements were considered superior to any other vessel on this route.

Although the order book was thinning somewhat with few enquiries coming in, the yard was busy with another large four-masted sailing barque, one of the largest ever, to be named *Crown of Austria*, Yard No 113. She was an order from Robertson, Cruickshank & Co. of Liverpool. At some 3,137 tons and with a length of 318 feet, beam of 45 feet 3 inches, and depth of 28 feet moulded, she had a carrying capacity of 4,800 tons, which would allow her to compete with the foremost steamers of the day. But the unfortunate *Crown of Austria* was wrecked on the Brazilian coast during her maiden voyage from Cardiff to Rio de Janeiro. R. & F. would build and launch a total of eight ships during this important year, and none would be more impressive than the large vessel ordered by the King of Siam. No ordinary steam yacht was this ship. She was large, at around 2,250 tons, and would also be armed with some 14 guns – almost like a small armed light cruiser. This magnificent armed steam yacht, *Maha Chakrkri*, Yard No 114, was named for the family name of the King of Siam. She was launched on 27 June 1892. This was perhaps the most important and spectacular launch ever seen in the port of Leith, attended by a whole heap of wealthy and well-to-do people, some representing the king and the government. Amongst the crowd were no less than four princes and a marquess, and many distinguished guests from as far away as Australia and Japan.

The *Maha Chakrkri* was cleverly designed with several special features including a double bottom that extended her full length to allow water ballast to be used when required to enter the many rivers of Thailand. Equally, she had pumps designed to expunge water to give the vessel a shallow draught and enable her to cross the bar of the river. She also had a projecting ram bow and an elliptical stern, with two funnels and two armoured gun emplacements on top of her masts forward and aft. Her armaments were completed with barbette mountings either side to take the latest breech-loading large calibre guns.

She was 300 feet long overall, with a beam of 40 feet and a depth, moulded to her upper deck, of 20 feet. Her two triple-expansion engines powered her twin screws, and she was supplied with electric lighting throughout, plus air conditioning units to the king (and his extended family's) private quarters. To massive cheers and many toasts, the *Maha Chakrkri* slid easily down the ways into the water at just past three in the afternoon, while the workmen looked on, no doubt wondering about the work still to be done and the dwindling order book in the shipyard. R. & F. had launched the much smaller steam yacht *Fauvette* the month before, and even though the outfitting on the *Maha Chakrkri* would take another two months, there were only three more ships to build that year; already there was rumour of layoffs and reduced working hours. The orders were for another screw steamer for the Currie Line; another very fine and quite large steam yacht; and then a special cable-laying ship that would take R. & F. to the end of the year.

Away from the pomp and ceremony of the launches, repairs remained the bread and butter of the shipyards. Companies such as Menzies & Co., which had built the *Sirius* some 70 years earlier, were heavily involved in repairs, and the Marr Brothers, amongst others, were busy maintaining wooden fishing vessels to keep the huge herring fishing fleets going. However, these were also very innovative times in maritime history, and Leith, as ever, was at the forefront; one such example was in using aluminium to construct boats. The following extract is taken from the Shipbuilding & Engineering magazine of 1892:

> Mr William Wells of Commercial Street Leith has recently patented and is now building a boat which is claimed is practically unsinkable and self-righting.
>
> The cruising yacht being built for a client in London will be built of a new white metal, supplied by the Alliance Aluminium Company of London. Owing to the peculiarity of this metal it will not be painted but polished. The fairness of the hull will provide for greater speed, while the boat at 32ft in length will be 35% lighter than one built from wood or steel.

His company went on to build many such yachts; in fact the following year was so busy he stopped taking new orders.

It was this year, 1892, that R. & F. Shipbuilders and Engineers officially became Ramage & Ferguson Ltd. The new company was set up with a capital of £60,000 in 500 shares. The shares were purchased by Richard Ramage and John Ferguson along with five others, none of whom were related to either Ramage or Ferguson.

Ramage and Ferguson were the new managing directors, and remained so until the resignation of Richard Ramage in July 1913. It was on his resignation that Henry Robb arrived in Leith to take on the position of steelwork manager at Ramage & Ferguson. Richard Ramage was to pass away seven years later, in July 1920, at the ripe old age of 86.

The ship building continued. SS *Minorca*, Yard No 116, was launched in August 1892, the eighth ship built to order for the Currie fleet; she was powered by a triple-expansion engine, also built by Ramage & Ferguson. She was to be another casualty of the First World War when she was sunk in 1915 by a German submarine with the loss of 15 of her crew.

Meanwhile, the rumours flying amongst the men with regards to the lack of work would soon turn out to be true. Already the Boilermakers Union had received notice that wages were to be reduced, and now an overtime ban was introduced in the yard. With the launch of the magnificent steam yacht *Valhalla*, Yard No 117, also in August, the men were put on notice that the working week would be reduced to eight hours (from ten) per day, and four hours on Saturdays, for the foreseeable future. While this was a blow, it was preferable to being laid off from work altogether; so the men agreed to the reduction in hours worked.

By October, it was further announced that anyone who came under the title of engineer or iron founder would also have their wages reduced. Ultimately, anyone earning 22 shillings or more per week would lose a halfpenny per hour, while anyone earning less than that would lose a farthing per hour.

This lack of orders was endemic all over the country, and it led to inevitable strife in the workforce, affecting the yards all along the coast of the Firth of Forth. The platers at Grangemouth shipyard went on strike, followed by some 800 riveters at Alloa. While the firms involved in dispute thought it acceptable to bring in apprentices to do the work of the tradesmen, that move only made the situation worse. This was an era when the employers were known and referred to as the masters and the workforce as the servants.

What work there was continued with the cable ship *Norseman*, Yard No 118, which was built and launched in November 1892 for the Western and Brazilian Telegraph Company. She would be the final vessel launched in this year of change and unrest. *Norseman* was fitted out with cable machinery, two cable tanks and twin bow sheaves by Telcon. Transferred to the Western Telegraph Company in 1899, she was renamed CS *Norse* in 1904 and sold in 1907. After a further five owners and name changes, she was scrapped in 1925.

In 1892, Ramage & Ferguson Ltd was still listed in the top 50 shipyards, measured by tonnage produced: 10,219 tons built, spread over its seven ships. The shipyard with the most tonnage produced the same year was Harland & Wolff in Belfast with 60,614 tons spread over a total of 14 ships built. While Ramage & Ferguson's output was fair, it was well below its 1891 total of 16,293 tons, even though its 1892 total included its four large sailing ships – as already mentioned, the largest ever built on the Forth – which contributed largely to the greater tonnage total.

The forecast for 1893 was not great, either. Of all the Leith shipyards, only Ramage & Ferguson had any building on the stocks, and there was work outfitting vessels already in the water – but when these tasks were completed the men would be paid off.

In the end, the yard had only two more – magnificent – steam yachts to build during the year, along with a very large four-masted sailing barque, the *Royal Forth*, of some 3,130 tons, and the steamer SS *Vala*, Yard No 122, launched in December 1893. The second of the steam yachts was the *Cleopatra*, designed by G.L. Watson, and she took an active part in the First World War.

As such, the relatively new limited company did not do so well in 1893, but little was predictable in the shipbuilding world and the next year proved much better. The first launch from the yard, in the spring, was the very fine-looking steam yacht *Earl King*, Yard No 123, designed by St Clare John Byrne. SS *Vana* (sister ship to the *Vala*). Yard No 124 came next. A steam-powered cargo ship of 1,021 grt, with a length of 219 feet, she spent most of her working days on the Baltic trade routes out of Grangemouth, where she was owned by J.T. Salvesen & Co. Then during the dark days of the Second World War *Vana* was purchased by the Ministry of War and was set up as a blockship at Yarmouth, complete with a shipload of explosives ready to be set off should that port come under threat of attack. In 1943, once the threat had receded, she was towed to Brancaster and anchored further out to sea to be used as target practice for a new type of cannon shell being used by the RAF against shipping – and very successful this was to turn out, as the RAF boys did not often miss! So when a north-westerly dragged her on her anchors to her present position near the beach, being full of cannon holes she sank where she remains today. (Information from the Brancaster Local History Group.)

Following the precedent of SS *Vana* came a further four magnificent steam yachts, all constructed between the building of three more steamers and a steam lighter, which was launched in September 1894. SS *Narova*, Yard No 126, was launched in April, eventually ending up as yet another war casualty. Renamed the *Gebil Yedid* and owned by Bland of Gibraltar, she was sunk by a German submarine in 1917 en route from Montreal to Gibraltar, luckily with no loss of life.

*Not much of SS Vana is left, but the wreck still attracts visitors, sometimes catching people out; the tides are treacherous in this area, as the warning signs point out. (Brancaster History Group)*

The aforementioned steam yachts were launched as *Zeta*, Yard No 125 (March), *Elian*, Yard No 127 (April – just two weeks after SS *Narova*) and *Ellida*, Yard No 128 (May). *Ellida* was an order for the owners of the local Currie Line. Some records show she was for Sir Donald Currie while others show she was for James Currie. Either way, she became the flagship of the Royal Forth Yacht Club – owned by James Currie – and her build quality encouraged Sir Donald Currie to commission a much larger steam yacht from the firm just one year later.

The *Ellida* was followed by the launch of the much larger steam yacht – and some would say one of the most beautiful – *La Belle Sauvage*, Yard No 129. She was an auxiliary steam yacht of some 531 tons, built for J.B. Robinson, a fine-looking vessel that would end up in French hands as a well-known sight on the yachting scene. The two smaller steamers built were sister ships, ordered for Hodge Daout Tarkou & Frankier Frères, and were launched as SS *Georgios* and SS *Sofia*, Yard No 130 and Yard No 131, respectively, with the construction of SS *Saint Ninian*, Yard No 133, taking the yard into the following year.

The *St Ninian*, a passenger/cargo steamer, was launched in February 1895, an order for the North of Scotland, Orkney & Shetland Steamship Navigation Company; she should not be confused with a similar vessel built the year before on the Clyde. A superb general finish over her cabin accommodation for 180 passengers ensured the ship was popular on the northern route from Aberdeen to the islands. In addition, she exceeded her required service speed during her sea trials with a speed recorded over the measured mile of 13½ knots to the satisfaction of builders, surveyors and owners. She would go into service very quickly, in April 1895.

It was at this point, with a fair amount of work under construction, that Ramage & Ferguson had to face the fact that all was not well with its shipwrights. They had decided to strike; 39 of the workers walked out over the issue of men wanting to reduce the amount of caulking done per day (from 200 feet to 120 feet) whilst still receiving the same amount of pay. They were soon joined by some 33 apprentices, which meant that no caulking would be carried out at the yard while the dispute was ongoing. Nevertheless, apart from this and other occasional strikes, the shipyard of Ramage & Ferguson was now considered a very prestigious builder of fine steam yachts. Some of the builds might have struggled to make any profit at all, but the prestige and goodwill engendered by building such vessels could never be calculated in monetary terms alone.

The early part of each year was known as the fitting-out season for steam yachts, as that was when they usually came back to the builders for overhaul before going out for the following yacht season. After some 50 steam yachts had been built at Leith, the port was seen amongst the wealthy owners as *the* place to go for overhaul and upgrades to their prized assets. Thus, in the spring of 1895 the yard was kept very busy with such work, as well as the construction of two brand new steam yachts.

These were built to complement the growing fleet of the Currie Line: SS *Nubia* and her sister ship SS *Corsica*, both ships of around 1,100 tons. The *Nubia* was built to carry coal on the company's Grangemouth–Hamburg route, and was launched as Yard No 136 by Miss Jardine of Edinburgh on 6 June 1895. SS *Corsica*, Yard No 138, was launched the following month. She was significant in that she was a new type of steamer complying with all the new rules and regulations of the Board of Trade. This meant she could be converted from the carrying of general cargo to that of a passenger steamer. SS *Corsica*, being the twelfth such order from Currie, meant that Currie was by far the largest customer of the Ramage & Ferguson shipyard up to this time.

The auxiliary steam yacht SY *Arcturus*, Yard No 140, designed by St Clare John Byrne for R. Stuyvesant of New York, would complete the construction of ships for the year, and another very fine and well-known steam yacht became the first launch of the New Year: SY *Iolaire*, Ship 141 (not to be confused with the *Iolaire* wrecked off Stornoway in 1919). This New Year's *Iolaire* was built to the order of Sir Donald Currie M.P., and would become the new flagship of the Royal Forth Yacht Club. (NB it appears that Sir Donald commissioned yet another *Iolaire*, built at Govan in 1902, which in 1939 was renamed HMS *Persephone*, and continued in service until 1948, when she was scrapped.)

1896 was to prove a somewhat barren year for the construction of ships at the Leith yard, with only two steam yachts, two steamers and a small steam launch built. SS *Rasona*, Yard No 142, was launched in January, two weeks after the launch of SY *Iolaire*. She was an identical vessel to SS *Narova*, previously ordered from Ramage & Ferguson Ltd by Thomas Cowan of Grangemouth. The first steamer built that year was a twin-screw yacht designed by Dixon Kempt for Baron Baretto, and she was to be the fastest steam yacht for her size, and the best equipped. She was followed a month later by the second steamer, the twin-screw SS *Frontier*, for the United Boating Co. The last build of the year, a small twin-screw steam launch named *Cooljack* was then launched.

1897 saw an upsurge in work, with Ramage & Ferguson Ltd building nine vessels, starting with the Cox & King-designed SY *Rosabell II*. Then came SS *Vestra*, Yard No 147, an order from J.T. Salvesen & Co., the first of only two steamships to be built that year. She was launched in January, the second steamship, named SS *Ronan*, Yard No 153, following in September. SS *Ronan* was an order from Geo. Gibson & Son. The rest of the year was taken up with the construction of steam yachts (and, incidentally, the construction of a grain elevator). A couple of the steam yachts would go on to become very well known and have interesting histories; their details will be given in another book I am currently writing on the steam yachts built at Leith.

The first of the prominent steam yachts, SY *Gunilda*, Yard No 149, was built for the oil tycoon W. Harkness, and she was to become the flagship of the New York Yacht Club. She was designed by Cox & King and launched in April 1897. The other well-known steam yacht, SY *Keithailes*, Yard No 152, followed in May. Ramage & Ferguson also built and launched the steam tug *Gaviao*, Yard No 151, during a relatively busy year.

1898 would see the construction of the largest steamer up to that time at the Leith shipyards – SS *Cathay*, Yard No 156, at around 4,112 grt. She was an order from Anderson & Co. as ship's managers for the East Asiatic Co. As mentioned in an earlier chapter, SS *Cathay* collided with SS *Clan MacGregor*, which came off worst, sinking off the coast of St Vincent. SS *Cathay* was to go down in 1915 after hitting a mine.

The steamer *Natuna*, Yard No 157, was launched in April 1898 to an order from the East Asiatic Co., Copenhagen. She was first sold on to the famous Norddeutscher Lloyd shipping line of Germany in 1900, and then sold on once more in 1912 to the Sarawak & Singapore Steamship Co. This remarkable ship of some 764 grt would survive both world wars and be sold on again a number of times to finally meet her end as SS *Irene*, owned by an Argentine shipping company; she was wrecked on the coast of Brazil in September 1958. Two of the companies involved in owning her must have been so impressed with her build that they would return to the Leith yard of Ramage & Ferguson for more orders: the East Asiatic Co., which would return to the same yard a few more times requiring much larger ships; and the Sarawak Steamship Company.

Another notable ship build during this very productive period was a ship ordered by William Thomson & Co. (later, the Ben Line). She was SS *Reval*, Yard No 158. Launched in July 1898, she survived the First World War and stayed with the company for some 26 years before being sold on to the Ulster Steamship Co. in 1924 and renamed SS *Dunmore Head*.

The final two years of the century would see the Ramage & Ferguson shipyard become synonymous with steam yacht building. All the very best designers would entrust their designs to be built at Leith, resulting in some very famous yacht names launched. For example, *Surf (I)*, Yard No 159, designed by Cox & King and often mistaken for another SY *Surf* built later by Hawthorns Ltd. Then there was the magnificent steam yacht, *Shemara*, designed by the same company, and launched as Yard No 165 in January 1899.

One order clearly illustrates the downturn in shipbuilding around the turn of this century; commercial ship orders being few and far between, the company managed to secure an order for the build of three fishing vessels, a type of ship not normally built by large shipyards – but any order was better than none. They were for a new local company, Newhaven Trawlers Limited; the first of the two was constructed right away and the third was to have her keel laid as soon as a berth was available. The first of the 105-foot-long trawlers to be launched was FV *Aurora* in March 1899, just after the steam yacht *Lady Gipsey*, Yard No 163, had been launched.

*Golden Eagle*, Yard No 166, another of the fine G.L. Watson designs, was launched one week later, and the *Aurora*'s sister ship, FV *Aries*, Yard No 165, was launched on the same day from the same slip.

The next vessel on the stocks at the shipyards of Ramage & Ferguson Ltd was another order from Currie. This, at 1,081 grt, was to be the largest ship built in 1899, and very welcome to the local workforce. She was a steel screw steamer built for use in the company's general cargo trade; launched on 24 May 1889, she was 225 feet in length with a beam of 33 feet 3 inches with a moulded depth of 16 feet 11½ inches. Her engines were of the triple-expansion type, with cylinders of 18½ inches, 30 inches and 49 inches diameter by a 33-inch stroke. The steam for her engines was supplied by a large single-ended boiler, working up to 165 lbs pressure, to be fitted by the builder. She was built to the highest class at Lloyd's, and as she went down the ways she was named *Elba* by Mrs Crawford of Trinity, Leith. SS *Elba*, Yard No 162, was in fact the 13th steamer to be launched for Currie & Co by Ramage & Ferguson Ltd. While on her sea trials carried out the following month, on 17 July 1899, *Elba*'s engines powered her through the water to achieve 10 knots on the measured mile, much to the satisfaction of her new owners.

Jas Currie & Co. Ships was managed by, and part of, the Leith, Hull & Hamburg Steam Packet Co. Ltd, and the *Elba* was still in the service of this company up to and during the First World War. In keeping with most merchant ships of the British Mercantile Marine, she was armed with a couple of guns to try and protect herself against the dreaded U-boats of the time. She survived her trading until April 1918, when she was torpedoed and sunk by UB-103 with the loss of ten of her crew. (For more on the sinking of SS *Elba* see Chapter 7).

Incidentally, it is estimated that during the First World War around 375 U-boats sank some 6,596 merchant ships, aggregating a total tonnage figure lost of around 12,000,000 tons – and that's just the ships. The loss of men who sailed on these vessels was also criminally high, at more than 14,500 during the four years of this terrible war.

For a short while, another war, the Transvaal (Boer) War, under way in South Africa, was adversely affecting the ordering of new ships on the Clyde, Leith and elsewhere. As a result, the *Port Maria*, Yard No 171, was not launched until December 1901. However, the slowdown in ordering did not last very long, as the British government required many steamships to transport an army and its equipment to South Africa, a situation that the prospective ship owners were not slow in taking up. Such was the perceived demand that shipyards had to begin to advertise for many squads of steady riveters; the incentive was long-term employment.

At the other end of the scale were the so-called banana boats. Yes, they were built at Leith. In fact they were amongst the very first such ships, with two others built on the Clyde and a third at Dundee, for a fledgling company transporting the still-exotic fruit from the West Indies to British shores.

During the last decade of the century, the shipyard of Ramage & Ferguson Ltd built eighty ships in total, an average build of eight ships per year: one ship was being built and launched on each of the company's four building berths approximately every six months. This is phenomenal considering that so many of these ships were some of the finest large private yachts ever to sail the seven seas; their workmanship would always require just that bit more time than the average vessel.

The penultimate order of 1899 was a barge constructed and launched for the Leith Dock Commission, and then the last vessel of the 1800s was a steam tug named *Panther*, perhaps fittingly as she was on order for the African Boat Co. She was launched into the Firth of Forth on 1 November 1899.

Amongst the ships built were four very large steel four-masted barques, all around 3,000 grt, and many more of the magnificent luxury steam yachts for which the shipyard was already well known. In addition there were the equally fine cargo steamers, such as SS *Moscow* for the Ben Line shipping company, and a steam tug, *Solway*, for the North British Steam Packet Co. Ramage & Ferguson Ltd also constructed 42 steamers. Some were for well-known local owners (Ben Line; Currie), others were for J.T. Salvesen and a range of companies near and far. With such an impressive record, there is little argument that the firm of Ramage

*An original stamp, dated 29 October 1897, taken from a Ramage & Ferguson drawing signed by Alex Ramage, who joined his father in the company around 1885. The same shape of stamp was used by Hawthorns & Co., while later the stamp with the name of Henry Robb Shipbuilders & Engineers would be seen on all drawings released for production work in the yard. (Author's collection)*

& Ferguson had well and truly arrived at Leith and deserved its place as the premier shipbuilders in the port.

# J. CRAN & CO, SHIP-BUILDERS AND ENGINEERS

The enterprise that was to become the Cran & Somerville Shipyard apparently started out at the foot of Leith Shore some way along from where it was to end up in the Victoria Yards. The firm was originally called John Cran & Co., but it was never a limited company until the formation of Cran & Somerville Ltd, sometime in 1917.

John Cran was from Aberdeenshire and he, along with his brother-in-law, a marine engineer from Glasgow by the name of David Donaldson McLellan, purchased the Britannia Inn and Stables, at Timberbush, Leith, complete with yard, in 1883. This was primarily to build ships' engines and boilers, as there was not enough space remaining at the new Shipbuilder's Row. When the firm of R. & F. Shipbuilders had finally concluded its lease of the land that was to form the Victoria Shipyards, Richard Ramage and John Ferguson realised that they themselves did not, or could not, use all of it, so they wisely decided to sub-lease part of the land. This was just the opening that John Cran had been looking for; the lease of this waterfront land would enable his fledgling company to begin the building of ships. Later, John Cran moved his engine-building works from the original small premises in Tower Street to the Shore, in larger premises which would eventually be taken over by the company of George Brown & Co., which is still in business there to this day.

A great many of the 130 vessels the firm built were tugs, such as the first to be launched, the *Swai Hijh*, which was of course Yard No 1. She was an order for J. & J. Bullock of Rangoon, launched in February 1883 (some five years after R. & F. had launched its first ship in 1878). Other tugs, such as *Fanny*, *Nea Omonoia* and *Huskisson*, were built, along with many stern trawlers and some cargo vessels.

In fact, it was engine building and the supply of engines, as well as ship repair and general engineering support work that would keep the company going throughout what is the natural cyclic way of shipbuilding. John Cran supplied steam engines to many of the local smaller builders of ships around Leith; there were many who would be building fishing trawlers and the like for the booming fishing industry around the British Isles. Small yards such as

J.M. Kenzie & Co. Boat Builders of Leith, which built FV *Evelyn*, had her engines supplied by John Cran, along with FV Fulmar, a wooden steam-powered fishing vessel. In the year 1885 alone, we see that John Cran & Co. supplied engines for the following fishing vessels: FV *Alice*, FV *Malektijar*, FV *Osprey*, FV *Petrel*, FV *Venture*, FV *Bull*, and FV *Merlin*. Along with the aforementioned *Evelyn* and *Fulmar*, and also FV *Perseverance*, all of these received compound engines.

In general, the years 1884–1885 had not been great for shipbuilding, so the supply and build of engines was a very good sideline. Such was the dearth of work on the Clyde that many men were out of work and facing destitution, if not starvation, with little or no financial help save that of the meagre amounts passed out by the various unions and charities of the day. The men and their families could only wait and hope for better days – or move overseas if the opportunities arose.

While the Cran shipyard tended to specialise in the building of tugs and fishing trawlers, it was to become better known for building many fine steam tugs, such as the *St Rollo*. Launched on 25 August 1888, she was of iron construction and built for the well-known lighterage company on the River Thames run by a Mr J. Constant. Relatively small in size, she had a length of 60 feet and beam of 14 feet, with a depth of 8 feet, and was powered by a compound surface condensing engine supplied by the builder. The *St Petrel* was a subsequent tug built for the same London-based owners and launched on 6 September 1888. Her engine was also supplied by John Cran & Co.

John Cran & Co. built only two vessels of note during 1892, the first being the steam tug *Harbour Light*. The other was the *Lady Alice*, a small water boat of around 50 feet in length, with a beam of 13 feet and a depth of 5 feet 8 inches. Once constructed and bolted together in Leith, she was then marked off and all her constructed parts disassembled to be shipped overseas to the port of Zanzibar where she would be rebuilt and launched on site. She was to have powerful water pumps supplied so she could also be used in the port as a fire boat.

A couple of years later, the yard received another very interesting commission; the newly opened Manchester Ship Canal (MSC) had ordered two steel tugs. The first of these was launched on 11 July 1895, named *Minnie*. With a length on her keel of 58 feet, a 13-foot beam and 7 feet 3 inches depth of hold, she was propelled by engines built by the boat builder, producing some 120 hp.

The tug *Minnie* was followed on 7 August by her sister ship, the tug *Gwennie*, of the same size and power. As it transpired, the MSC was to be a very important customer for the future Henry Robb Yard. Also built during 1895 were a couple of steam-powered launches for foreign owners, along with another tug named *Venture*.

In the year 1899, although the small yard had not built a ship, it was busy with repair work and the building of marine engines; some six sets of engines were to be produced for trawlers, so the marine work was steady if not spectacular at a time of record production of tonnage around the rest of Scotland.

It is safe to say that while J. Cran was busy with tugs and fishing trawlers, the majority of tonnage in new building was carried out by its neighbour, R. & F. Shipbuilders. However, S. &

H. Morton was also a formidable competitor, building steamships in the range from 500 grt to 1,200 grt. The *Norna* has already been mentioned, and another significant steam cargo ship Morton built was the specially designed fruit carrier for trade between the West Indies and the United States. Built for an order from Harloff & Boe of Bergen, SS *Baracoa*, Yard No 59, was launched on 7 April 1890 with a length of 180 feet, a beam of 28 feet, and a depth of 12 feet 8 inches. She was driven by triple-expansion engines made by the builder. Yard No 60, SS *Morena*, was another steam cargo ship of some 1,292 grt, launched in August 1890, also from the shipyard of S. & H. Morton. She was an order for Currie, subsequently wrecked near Cape Ray, Newfoundland, in May 1907. Morton was also responsible for SS *Sterling*, Yard No 61. She was a triple-expansion passenger/cargo ship of some 1,047 grt. Launched in Sept 1890, she survived the First World War but was wrecked on 1 May 1922 at Seyðisfjörður, Iceland.

The year was to finish with the build and launch of two barges which were built using the shipyard's patented longitudinal and rather innovative system of construction, which gave much more strength and carrying capacity with less weight. It comprised shell panels of a 'Z' form so that they could all be riveted with a hydraulic rivet machine, saving much time in construction.

S. & H. Morton had had another order from T. Hughes & Co. of Liverpool for the construction of two iron hopper barges; however, circumstances halted their progress. On 1 December, Morton issued the news that it had suspended payments and at a meeting of creditors the estate showed liabilities of some £28,000 with assets of around £10,000. But even with this news it was expected that a settlement would soon take place and the yard would resume working as before. Meetings were set in motion and it soon emerged that although talks were ongoing, the yard would indeed continue under the same name, run by two of the previous partners and another investor from Edinburgh, well known in engineering circles. It was not a great way to be finishing any year and certainly not a good situation for all the men employed in the shipyard.

If towards the end of the 19th century you looked at an aerial view of the land that now contained the shipyards known collectively as the Victoria Shipyards, you would see starting from the west pier (on the right-hand side of the photograph in the introduction) the four slipways that formed the firm of Ramage & Ferguson Ltd. Adjacent to that was the firm of S. & H. Morton Shipbuilders and Engineers with three berths for building ships, and at the furthest inland point going east would be the two berths of J. Cran & Co. So there was a total of nine building berths all capable of launching ships directly into the sea, with D. Allan, who built trawlers, squeezing in there somewhere between them. To summarise, then, from around 1885 Leith had two shipyards capable of building ships in excess of 400 feet in length, with the smaller John Cran yard specialising in the building of tugs – and they could all build their own engines as well.

In the 50 years of building ships up to the turn of the new century in 1900, a total of around 280 ships were built by the three shipbuilders on the site, and this does not take into account the many fishing trawlers built by D. Allan. In addition, a very large number of marine engines were built by all three of the existing shipyards, numerous ships were repaired or converted, and S. & H. Morton also built some of its own patent slipways.

As is the way in shipbuilding, then, in just a few short years there had arisen an unprecedented clamour from owners old and new for ships, and so record amounts of new tonnage was being built – not just at Leith but at the Clyde, Dundee and Aberdeen shipyards as well. There was enough skilled labour in Leith to avoid any problems between the different builders; they were all, it seems, specialising in certain types of ships in order to avoid too much competition.

# FIVE: A NEW CENTURY, 1900–1910

As the world bade goodbye to the 19th century and welcomed the 20th, the Boer War had been going for a just over a year and would continue on taking lives until its completion in 1902, bringing the Victorian era to an end and heralding the Edwardian. The offshoot of the Boer War was a financial depression that was to last until 1911 – throughout the Edwardian era and into the first year of the reign of his son, George V – taking its toll on the working people during this decade, (although as ever it had comparatively little effect on the 1 per cent who virtually owned the world between them). It was in this small group of privileged people that the shipyard of Ramage & Ferguson found a very good market in the continued building of ever larger and ever more luxurious steam yachts.

There was still a great deal of unemployment in Leith during the Edwardian times. The shipyards were an ideal breeding ground for trade unionism, attracting men who wished only for a fair day's pay for a fair day's work. Meanwhile, the owners and managers (at the time still called masters) only wished for higher profits and a larger return for the shareholders.

It was a period hot in the pursuit of excellence in shipbuilding and the design of ships; boom years for shipbuilding and a time when the shipyards of the British Isles were to reach their zenith. The ten years before and ten years after the turn of the century were the real golden years for Britain as a nation: it not only ruled the waves but built the vast majority of the world's ships, too. However, as we have seen, the fortunes of the shipbuilding industry have always ebbed and flowed just as the seas its vessels populated, and the eventual decline of this mighty industry would be seen in the quote below from Norman L. Middlemiss in his excellent trilogy on British shipbuilding yards first published in 1995:

> Few industries can attest to the decline of Britain's political and economic power as does the near disappearance of British shipbuilding. On the eve of the First World War, British shipbuilding produced more than the rest of the world put together. But by the 1980s, the industry which had dominated world markets and underpinned British maritime power accounted for less than one percent of world output.

But that was yet to come. For now, the shipyard of Ramage & Ferguson was going from strength to strength, assisted with backing from the financial clout of the Ellerman Shipping Company. Ramage & Ferguson was now recognised as one of the premier shipyards for small to medium-size shipbuilding, not only in Scotland but throughout the whole of the British Isles.

The new century began with the build and completion of six ships. The steamer SS *Sappho*, Yard No 169, launched on 30 January 1900, was an order from the Bristol Steam Navigation Co Ltd for its general trade between Germany and Bristol. The *Sappho* was quickly followed by SS *Port Maria*, Yard No 171, a steam cargo/passenger ship of some 2,910 grt. She was launched in October 1901, part of an order of three other ships being built elsewhere in Scotland, and she would have the capacity for some 20,000 packages of bananas. Her accommodation serviced 25 first-class and 12 second-class passengers, and she was to be run by the Jamaica Shipping and Transport Company. This was a new company formed to provide a direct fruit and passenger service between the West Indies and the United Kingdom. The timetable scheduled steamers running a fortnightly service at an average speed of 15 knots between Kingston / Port Antonio and Southampton.

For good measure, a couple of very fine luxury steam yachts followed, culminating in the build of a large steamer for Ben Line: Yard No 174, SS *Bencleuch*. At 4,159 grt, she was a large vessel for the yard and was to give good service to Ben Line, surviving the ravages of the First World War and finally being sold on in 1920 to Hajee H. M. Nemazee, Hong Kong, and renamed *Tangistan*.

SS Bencleuch, *date unknown. (Vancouver Archives)*

Ben Line had some six or seven vessels with identical names, so the Bencleuch should not be mistaken for her namesakes. She was sold out in 1920 and quickly replaced by a vessel awarded to Ben Line as a war repatriation. This ship, originally named *Emden,* was also briefly given the name of *Bencleuch* before she herself was sold on to the Hapag Line in 1921 and reverted to *Emden.*

An interesting ship built in 1901 was the twin-screw tug *Helen Peele*, Yard No 176, built for the RNLI and designed by the man perhaps better known for his designs of luxury steam yachts, G. L. Watson.

Without a shadow of doubt, Watson is worthy of a significant digression here. He was a pioneer, one of the premier designers of his time, esteemed alongside greats like William Fife and Nathanael Greene Herreshoff. Over Watson's somewhat short lifetime he was responsible for 450 designs of yacht, boats and ships, as well as improving and designing new lifeboats. He was born in Glasgow in 1851, only 50 years after the first successful use of steam power in a ship using the William Symington patented marine steam engine. This was the steam tug *Charlotte Dundas*, which had carried out trials on the Union Canal just outside Edinburgh in 1801. While the *Charlotte Dundas* was towing her two laden barges along the canal, the population of Glasgow was only 67,000. When Watson arrived in 1851, the population had grown to around 400,000, an increase of 600 per cent in 50 years.

Watson was from a well-connected and wealthy family and had attended a private school in Glasgow. Rather than going to university, he had wanted to go into engineering; this was at a time when the great engineers were regarded as hero figures. So through family connections he was entered into the well-known and distinguished shipyard of Robert Napier, the previously mentioned father of shipbuilding on the Clyde. As a premium apprentice (meaning his family had paid for his training), he started out in the drawing office of Napier's esteemed shipyard, before leaving at 18 years old to complete his apprenticeship over the river from Govan at the A. & J. Ingles & Co. Pointhouse Shipyard. As a result of his experience in these two shipyards and with his natural talent, he was destined for greatness. He branched out and set up what was the very first dedicated yacht design company in Glasgow in 1873. The first of his designs was built at Leith, and was, perhaps surprisingly, very far removed from his later signature luxury steam yachts; an order was placed with the Leith shipbuilding firm of A.G. Gifford & Co., a well-known fishing vessel builder in the port, for a barge and cargo/passenger ship of 23 grt for the Loch Tay Steam Boat Co. But later the SY *Cleopatra*, Yard No 121, a steam yacht built by R. & F. Shipyards in 1893, was the first of many beautiful ships to be designed by Watson and his company to be built at Leith. He would have more built at Ramage & Ferguson (which is what R. & F. were to become) over the next 30 years or so. While many of his fine designs would also go to other shipyards (mostly on the Clyde) it is perhaps a testament to the renown of the shipyard of R. & F. that the great designer entrusted some of his work to that Leith shipyard. He was well-known as a perfectionist and would have been greatly interested in how his drawn creations turned out as real ships. Watson died in 1904, perhaps from overwork as this genius of a man was never idle. His company, now based in Liverpool, England, carries his name to this day.

But back to the ships themselves … The *Helen Peele* was a twin-screw tug designed by G.L. Watson to an order from the Royal National Lifeboat Institute, and she launched as Yard No 178 in June 1901. She was followed by the build of a couple of large cargo steamers, the first of which was SS *Bin Tang*, Yard No 177, an order from the East Asiatic Co. of Copenhagen. She ended up as a converted minelayer for the Imperial Russian Navy, and in 1917, with the Russian Revolution in full swing, she was scuttled in Archangel by the Bolshevik forces to prevent her falling into the hands of the incumbent British forces.

Launched just a couple of months later, in September 1901, was an order for the Sarawak Steamship Company (which would return to Leith for ships later in the century). She was named SS *Rajah of Sarawak*, Yard No 178. There was a further three-ship order from Currie & Co.; the first two were SS *Scalpa*, Yard No 186, launched in May 1902, and her sister ship, SS *Staffa* (just two gross tons lighter, at 1,008 grt) following in the August of the same year.

The final order was the cargo steamer SS *Vienna*, Yard No 188. She became somewhat infamous in the British Isles in August 1914 when, renamed SMS *Meteor*, she was seized as a prize while in harbour at Hamburg when the First World War broke out. To take advantage of her unmistakably British appearance, the Imperial German Navy decided to convert her into an auxiliary cruiser and minelayer. She was moved to the Kaiserliche Werft (Imperial Shipyard) in Wilhelmshaven, where she was equipped with two 88 mm guns and two machine guns, and had minelaying equipment installed with a capacity for 347 mines. As the *Meteor*, she was commissioned in May 1915 under the command of Korvettenkapitän Wolfram von Knorr and tasked with minelaying to harry shipping in the North Sea. She was ultimately apprehended by a British light cruiser in 1915, but her German crew managed to scuttle her before the sailors arrived to capture her and return her to British hands.

SS Vienna, Yard No 188 (*University of Glasgow Archives & Special Collections, Currie Line collection, GB 248 UGD 255/5/7/30*)

SS *Farraline*, Yard No 191, the next cargo steamer, was built and launched in June 1903. She was ordered by the London & Edinburgh Shipping Line, and at 1226 grt she was a standard-type cargo vessel which met her end in November 1917, sunk by a German submarine with the loss of one life. In the following year came an order from the War Office for the build of two submarine mining screw steamers which today would be described as submarine chasers. Although the threat of submarines would only be fully realised later in that decade, it was a sign that the Admiralty was not altogether blind to those weapons and their potential. The two were to be named *Miner 17* and *Miner 18*, the first of many vessels built for the Admiralty at the shipyards of Leith.

Notable within the luxury steam yachts of 1905 was SY *Minona*, another design from G.L. Watson, and a vessel that is still around today. She has had a few owners, the best known of whom were the actors Richard Burton and Elizabeth Taylor, who changed her name to SY *Kalizma*.

*The steam yacht* Minona, *Yard No 204, in this fantastic photograph. The caption on the reverse of the photograph reads: 'Colonel Clark of Paisley with party aboard yacht* Minona, *passing through the Caledonian Canal, leaving Dochgarroch Lock, Sept 3rd 1937.' (© G.L. Watson & Co. Ltd Archive)*

In the same year, amongst the build of the yachts was the build of SS *Melrose* for Geo. Gibson. She met her untimely fate in 1940 with the loss of 17 of her crew, killed when she struck a mine

The following year, a vessel was built – one of the first of many – for a customer on the other side of the world. This was the Hunter River Steamship Company of New South Wales, Australia. The order was for a screw steamer named SS *Hunter*, Yard No 208, of some 1,840 grt. Surviving the First World War, she eventually succumbed to American bombs during the Second; while operating as a Japanese ship, she was spotted off Vietnam and sunk.

Yet another significant shipbuilding order was soon to be won by the yard of Ramage & Ferguson. This was for the *Terawhite*, a steam tug ordered by the Union Steamship (U.S.S.) Co. of New Zealand – the first of many such ships to be built first at Ramage & Ferguson and later at the Henry Robb Shipyard. The U.S.S. Co. would return to the Leith shipyards for a further 50 years or so to order ships, and would become one of their largest customers. The *Terawhite* might have been a small ship, but she had huge significance, as the company must have liked what it saw; it returned with a two-ship order for steamers to be used in New Zealand.

In 1907, a couple of interesting ships were ordered for the Leith, Hull & Hamburg Line, managed by the Currie Line. They were SS *Oder*, Yard No 216, and SS *Alster*, Yard No 217. Both were named after German rivers, and both were steam-powered cargo ships of 965 grt. They were launched in February and March 1909 respectively. SS *Alster* was sunk in the First World War, somewhat ironically by a German U-boat; she was torpedoed by UB-62 and sunk in 1918. Her sister ship, SS *Oder*, was bombed and sunk by German planes in September 1942.

SS Alster, Yard No 217 (*University of Glasgow Archives & Special Collections, Currie Line collection, GB 248 ACCN3877/5/2*)

Many more fine steamers and steam yachts were built between 1907 and 1911, by which time the country was beginning to show signs of coming out of the economic recession which had lasted since 1902. The year of 1907 ended with the launch of one twin-screw steam yacht, *Liberty*, and 1908 began with the launch of another, the *Iolanda*, Yard No 213 and Yard No 214, respectively. Then SS *Karuah*, Yard No 215, was another order from the Hunter River Steamship Company of Australia; she was a single-screw steamer of 399 grt, launched at Leith in July 1908.

SS *Slemish*, Yard No 218 (ON 124677), was an order from the Shamrock Steamship Company of Larne, near Belfast in Northern Ireland. She was a single-screw steamer powered by a triple-expansion engine giving her a speed of around 11 knots. She was launched in June 1909 with a grt of 1,536. She survived the First War only to meet her end in the Second. Fitted with one gun forward and one aft, both manned by gunners, she was classed as a DAMS. (In the Second World War, they were called DEMS.)

While on a voyage from Hull to Cherbourg carrying coal, the *Slemish* was sunk on 23 December 1944 about 40 miles north of Cherbourg. Six of her crew and one gunner were lost. The explosion, whether from hitting a mine – or as some records wrongly suggested – a torpedo from U-772, was just aft of midships and the ship broke in two, taking only a few minutes to sink. A nearby U.S. Navy vessel, USS *PC-533*, was first on the scene to render assistance, and managed to recover eleven crew members, along with her other three gunners, landing them at Cherbourg the following day.

The Trinity House Lighthouse tender THV *Argus*, Yard No 219, of some 653 grt, was launched in October 1909. She was powered by two triple-expansion engines driving her twin screws, and the following details some of her wartime service during the First War. This text was first seen in a Trinity House newsletter:

> From 2 December 1916 until the end of hostilities one of the steamers was continuously employed on the Dover Barrage. An interesting operation was reported, date not known, when Vestal, working on the Northern Barrage between Orkney and Norway, moored buoys at up to 160 fathoms secured to sinkers by four-inch wire moorings. As part of the continual change of buoyage required by the Admiralty from 28–30 October 1914 *Irene*, *Alert*, *Stella*, *Satellite*, *Mermaid* and *Warden* were engaged in removing 106 buoys in the Thames Estuary. From 20-22 September 1915, those vessels along with *Argus* and *Warden* moored the whole of the swept Channel buoys between Kentish Knock up the East Coast to Flamborough. Our ships were frequently fired on from the enemy occupied Belgian coast.

The THV *Argus* was to survive the First World War but not the Second.

# 1910-1914

Some five months after the *Argus*, the next vessel to be launched was another magnificent steam yacht named SY *La Resolue*, Ship 220. She was classed as an auxiliary steam yacht as she had both engine power and sails, and was built for a French sugar tycoon. At around 696 grt, she was a large and beautiful luxury ship, capable, with her sails and engine, of cruising far and wide. Many of the fine steam yachts rarely left their home port, or if they did venture out it was only for short cruises. After all, these signs of the super-rich were primarily to be seen and to show off how wealthy one was.SY *La Resolue* launched in February 1910 and was quickly followed by the build and launch of the single-screw steam cargo ship SS *Kingstown*, Yard No 221. She was powered by a triple-expansion engine producing around 81 hp, powering the 628 grt vessel through the water at around 10.5 knots. SS *Kingstown* (ON 128863) was another survivor of the First World War only to fall in the Second, when in 1941 she was attacked by German bombers. Although not sunk at that time, she would eventually sink while under tow about 6 miles off St Anne's Head.

The next small coastal steamer to be built and launched was an order for the firm of W.J.R. Harbinson of Larne. Named SS *Carnduff*, Yard No 222, and launched in June 1910, she was constructed with one full deck, part of which was iron, and she had an aft deckhouse with an open-to-the-elements wheelhouse. She was powered by a two-cylinder engine built by Ramage & Ferguson. Although a standard type of ship, she did have several unusual things happen to her during her long working life, and she was to run aground more times than, perhaps, her owners would have liked. Standard-type short sea traders of the day were very strongly built, and for such a ship to enter the harbour and be tied to the quayside while the receding tide left her high and dry on the mud at the bottom of the harbour was a common practice, as most small harbours were tide-dependent. Her strandings were to occur over a long period of time; the first recorded one was in 1934, in what is now known as Donegal Bay, and the last in 1950 when she ran aground in dense fog in Larne Loch in Northern Ireland; she appeared to suffer no great damage and was refloated the same day to continue her voyage.

SS *Carnduff* did, however, suffer a major grounding incident in 1946 on voyage from the Solway Firth with a general cargo bound for Larne. This time it was five days before they could get her floating again. With damage to around 20 shell plates, 15 had to be replaced and the rest were faired back into the shell. Fog around the coast of the British Isles was always a potential challenge for any master, and five groundings over a life of some 45 years appeared acceptable to her owners, as she stayed with the same company all her working life. Registered at Lloyd's and classified as +100 A1, her official number was 55602, her port of registry Belfast and she flew the Red Duster to show she was British. At a gross tonnage of 257 tons, the Carnduff was just one of many, but that she survived the terrible losses of both world wars was quite an achievement. She was finally scrapped at Belfast in 1955.

SS *Estrellano*, Yard No 223 was next to be built and launched by Ramage & Ferguson, Ltd, in 1910. She was a cargo steamer of 1,161 tons with a triple-expansion engine to power her along at 10.5 knots, and her official number was 131302. On 31 October 1917 on a voyage from Oporto to London with general cargo, she was torpedoed and sunk by the German submarine UC-71, 14 miles from Île du Pilier. Three persons were lost from the crew of 27 on the ship. She was owned at the time of her loss by Ellerman Lines, Ltd, Liverpool.

The giant shipping company that was Ellerman Lines was buying into the Ramage & Ferguson shipyard in a big way, so much so that it would soon have a major role in decision-making, with representation on the board of directors and a large part of the shareholding. With each large share purchase by the Ellerman Line the influence of Richard Ramage and John Ferguson was being diluted, probably leading to tension in the board room.

A very significant order, for SS *Kanna*, was won in 1911 from the Union Steamship Company of New Zealand (which had commissioned the steam tug *Terawhite*). *Kanna* might have seemed like just another cargo steamer belching smoke from her stack, but she led on to some 25 other ships built for this distant company. She survived the First World War and was eventually sold off by the U.S.S. Co of NZ to end up as a Japanese-flagged ship during the Second World War, when she was sunk by an American submarine in October 1944; the USS *Snapper* sent her to the bottom, 450 miles south of Yokohama. SS *Kanna* was part of a double order of ships from the U.S.S. Co; the second ship, SS *Karamu*, was built the following year.

SS *Koopa*, Yard No 227, was built and launched in 1911. Based in Brisbane from December 1911, and running up to Redcliffe and Bribie Island and back, she was to become known as the Queen of Moreton Bay by the locals who used her services in this northern part of Australia. The *Koopa* had proved her seaworthiness on the handover trip out to Australia from Leith when she had ridden out two great storms, one in the English Channel and one in the Bay of Biscay.

She was requisitioned for service in the Second World War as *HMAS Koopa* (pennant KP) from September 1942 to January 1947 and used as a depot ship at Toorbul, Queensland, and later as a mother ship with Fairmile Motor Launches in the Milne Bay area of New Guinea. She returned to the Bribie Island service in February 1947, then was acquired by the Moreton Bay Development Company, which went into liquidation in 1953. The *Koopa* was later taken off the Brisbane–Bribie run and then retired in 1960.

Looking at the ship in the photograph it is evident there are direct resemblances with later Henry Robb vessels built for passage on distant rivers of the world, although some of the future ships would be built at Leith and then disassembled and rebuilt *in situ* at the country where they were required, a sensible way around the stability criteria required for a shallow-draught river ship going deep sea.

Sea voyaging, even with the newfound and much better steam engines, was still a very hazardous line of work to be involved in, and some of the statistics for 1911 show just how perilous this work was. The summary of vessels totally lost, broken up or condemned etc, as published in Lloyd's Register for the year 1911, showed that around 888 vessels, amounting to some 884,843 tons of shipping – and not including any vessels of less than 100 tons – were lost during the year. Of these, 427 ships totalling some 619,752 tons were steamers, while a further 461 were sailing ships, totalling some 265,091 tons.

*Spectators watching SS* Koopa *on the Brisbane River. (State Library, Queensland, Australia)*

Not all losses were due to mishaps. Many ships were broken up as being surplus to requirements, and even more were just condemned as death traps; those losses, which amounted to around 30 per cent of ships, were removed from the Register in this way. The rest were wrecked by stranding, sinking and such, with an approximate 50 per cent split between steamships and sailing ships, showing that the sea cares nothing for the material of build or how the ship is powered.

As the number of lost ships in the 1911 Register was but a small proportion of the mercantile total of ships on the planet, and as it was similar to losses from other countries, and as the number was slightly down on the previous year, this number of losses was considered reasonable. It is also interesting to note from the official figures given that they only deal with lost tonnage; no mention is made of the lives lost.

The above losses were, of course, from the year prior to the 'unsinkable' *Titanic*, which was lost in April 1912. (Incidentally, the claim as to whether anyone from the builders actually said she was unsinkable can never be proven.) It is interesting to note that this sinking occurred 100 years after the first practical steamship, SS *Comet*, had been built and used on the River Clyde to carry passengers.

In this year of 1912, some of the first ships were being built powered by diesel, a new way of supplying power for vessels that would be taken up in a big way at Leith some ten years later by Henry Robb.

The lead ship of a twin order for ships by the North of Scotland, Orkney & Shetland Steam Navigation Co., launched in March 1912, was to be named *St Magnus*, No 229. Then

SS *Bernicia*, Yard No 232, was launched in November 1912, yet another order – the seventh – from the Leith, Hull and Hamburg line managed by Currie & Co., with an overall length of 220 feet 6 inches and breadth of 32 feet, depth moulded to the main deck of 13 feet 6 inches. She was built for the Leith–Newcastle passenger service, replacing one of the ships of the line that bore the same name but had recently been sold. SS *Bernicia* was built for 300 cabin passengers with some 200 steerage, and was to include all modern conveniences of the time.

SS *Palmella*, Yard No 233, was built and launched in December 1912 for the Ellerman & Papayanni Line. This ship was constructed in such a way as to be able to create a much larger unsupported span around her hatches. This was done by the use of the longitudinal pillar system without any side stringers or hold beams, and that meant that much larger loads could be craned into her holds. She was also designed with a cruiser stern which provided large deck space in the poop, which was also to prove beneficial when the ship was in a heavy following sea – a common enough occurrence in her trade route from London to Portugal through Biscay.

This was a way of building that the future Henry Robb Shipyard would use to great advantage in the design and build of ships that required very large hatch openings. The *Palmella*, herself, though, was built at an unfortunate time in world history, with the First World War just a year away; and, sure enough, she was to be sunk in 1918 when she was torpedoed by UB-92 near South Stack, with the unfortunate loss of 28 lives.

Into the new year of 1913, the first vessel to be launched in March was SS *Wendy*, Yard No 234, ordered by George V. Turnbull & Co. of Leith for its trade to the Spanish peninsula. SS *Wendy* was to end up as a Christian Salvesen ship. Then the second ship of the previously mentioned two-ship order by the North of Scotland, Orkney & Shetland Steam Navigation Co. followed the *Wendy*. Named the *St Margaret*, Yard No 235, and launched in April 1913, she was another fine passenger/cargo steamer with accommodation for a large number of passengers and the capacity to carry around 350 tons of cargo. She was 215 feet in length, with a beam of 31 feet and a moulded depth of 16 feet 2 inches.

July 1913 saw the resignation of Richard Ramage, co-founder of the original Ramage & Ferguson partnership, and in that same year Henry Robb was recruited from the west to become its shipyard manager. Robb was known to both Ramage and Ferguson, the three of them at one time or another all having worked in the same shipyard on the lower Clyde: Denny's of Dumbarton. Robb would spend the next four years at the yard before striking out on his own, and if there was strife in the board room, that would soon pale into insignificance with the events soon engulfing the world in war.

The large steamship SS *Transvaal*, Yard No 237, at some 4,395 grt, and 375 feet in length, was amongst the last of the ships built before the war began in August 1914. She was another order from the growing East Asiatic Co. of Copenhagen for its trade to South Africa and Australia. This ship was in fact destined to end up surviving the First War and was eventually sold to Germany in 1927, renamed *Gerrat*. She was believed to have been scrapped in 1934.

*At the time SS* Transvaal *was the largest vessel to have been launched in the east of Scotland. Due to the angle the photograph in this postcard was taken at, it looks as though she is heeled over to port!*

A total of 68 ships were built by Ramage & Ferguson from the turn of the century to the start of the First World War. The shipyard was building an average of five ships per year on its four building berths. Of the 22 ships built from February 1909 to February 1914, some 11 of them were sunk in either the First or the Second World War, a chilling statistic: 50 per cent of the ships built doomed to die before their time due to conflict. The number of ships lost to enemy action is bad enough, but many more would also be lost due to the hazards facing ships and their crews in the days when navigation hazards were incorrectly charted and weather reports were somewhat haphazard, not to mention the poor training – or lack of training altogether – when it came down to the crews.

The company alone was responsible for building around 10 per cent of all the luxury steam yachts in the world from around 1880 to November 1914. Altogether a total just in excess of 100 luxury steam yachts were to be built at Leith – more than twice that produced by the more illustrious shipyards on the River Clyde during the same period. Many of the steam yachts would play their own part in the forthcoming war, SY *Banshee*, SY *Rosabelle (III)*, SY *Rovenska*, SY *Honor*, SY *O-WA-RA*, SY *Liberty* and SY *Ul* are some of the ships built between 1900 and 1911.

Richard Ramage could be proud of his own particular shipbuilding legacy, which was carried on at this location for another 71 years after the beginning of the First World War.

# J. CRAN & CO., SHIPBUILDERS AND ENGINEERS

J. Cran's specialism was in tugs, single-screw and twin-screw; if you had a requirement for a tug, this was the firm to approach. This was well documented in the previous chapter. From the turn of the 20th century to the start of the First World War, the company would build 55 vessels (tugs and steam fishing vessels) and have a thriving repair business, and an engine and boiler production line. One of its most interesting tugs in this time frame was the 201 grt FV *Herculaneum*, Yard No 51, launched in February 1905, which was to survive the war and was in fact renamed FV *Formby* in 1927.

Not to be outdone by its more illustrious neighbours, which were building luxury steam yachts by the dozen, the small firm of J. Cran completed some steam yachts as well in this first decade of the 1900s; one such vessel was the SY *Mereli*, launched in July 1906.

In 1907, the firm built and launched a couple of steam trawlers and a couple of steam-powered tugs, the first one named FV *Betty*, Yard No 59, followed by another trawler, FV *Bass Rock*. She was a 169 grt steam trawler that was sunk by bombing on 24 September 1940 off the Old Head of Kinsale. The German bombers based in Norway were regular visitors to the east coast of Scotland at this time, and they apparently regarded an unarmed fishing vessel as fair game. Her skipper and three of her crew were killed. They are commemorated on the memorial for Merchant Seamen 'with no grave but the sea' at Tower Hill, London

The following year, 1908, the firm would build another two tugs named *St Hornby*, Yard No 66, with her sister ship the *St Egerton*, Yard No 67, launched two months later, in March. MV *Solent* (one of a two-ship order) was a pilot auxiliary motor ketch; she was of particular interest as she was propelled by engines supplied by W. Beardmore & Co. of Dalmuir which used paraffin as fuel. Her engines ran without a hitch during her sea trials and she attained a speed of 7 knots average over the measured mile. She was only 49 tons. Built for Trinity House in 1910, she sank in December 1912 after a collision in the Solent with HMS *Dufferin*. Another tug built was ST *Bristolian*, Yard No 76, launched in February 1911. This small tug,

*The steam tug* Alexandra, *Yard No 108, was built in 1907 and typical of the many fine tugs that Cran was to build during this decade of the new century. (Author's collection)*

of around 174 tons, was to have a very long working life, surviving two world wars, and was only broken up in 1968, in Newport, Wales.

The next few years saw the build and launch of more tugs and a couple of small cargo steamers, the first named SS *Inniskea*, Yard No 88, launched in May 1912. She broke down on her maiden voyage off Tory Island, Northern Ireland, outbound from Glasgow, and while her crew managed to jump to shore safely after a tug towline had parted, she was later driven onto the rocks the next day, 12 October. A salvage team boarded her to find her hull pierced by rocks, and the following day she broke up in the teeth of a storm. Her sister ship, SS *Innislargie*, was launched from the yard in September 1912, Yard No 92, along with another two tugs, *St Lillian* and *St Vivian*. The yard also carried out a lot more dry dock and repair work than the previous year.

With the huge company of Alexander Towing Co. Ltd of Liverpool patronising John Cran & Co, the future was looking bright for that small yard, and it launched yet another tug for this company on 19 June 1913, the second of a two-ship order (the previous ship having been launched a few months earlier). The *Gladstone*, Yard No 96, was the 18th tug to be built with engines by the yard for Alexander Towing Co. Ltd. She was a steel screw tug/passenger tender of some 211 tons with powerful engines also made by the builder, to produce over 900 hp.

Two more tugs were on order for the Antofagasta and Bolivia Railway Company, the first of them being launched as the *St Berta*; the final two ships launched, just as the First World War was beginning, were two steam fishing trawlers: FV *Craig Island*, Yard No 100, in August 1914 and FV *Anwoth*, Yard No 101, in December. Both vessels survived the war but would eventually succumb to the sea, the *Craig Island* being wrecked near Aberdeen harbour when she ran aground in 1922, and the *Anwoth* running aground in 1932 and wrecked on Sheep island, Mull of Kintyre, with all crew saved.

# THE SHIPYARD OF S. & H. MORTON

S. & H. Morton also continued to build ships during this decade, although a lot of the work that it found was in the repair of ships calling into Leith. The shipyard of course also continued with the building of large patent slipways for customers worldwide, and it was known to lease out some of its own three slipways, when empty, to enterprising firms engaged in the build of large barges and such. It was finding the competition difficult from the other two yards which, by the beginning of the new century, had both attracted very large backers of shipbuilding operations carried out in Shipbuilder's Row – now better known as Victoria Shipyards (due reference to the passing of the old queen of the same name after a record time on the throne of Great Britain and her Empire, much of which had been gained during her reign).

If we take the year of 1909, for example, the old shipyard of S. & H. Morton did not build a single vessel, although it was engaged in much other marine type work such as the building of marine slipway hauling engine and gear for Bombay, along with a large number of horizontal steam cargo winches, and many ship repairs and engine repairs. The lack of shipbuilding orders, though, was proving to be more and more difficult, and the old yard was in terminal decline, seemingly unable to secure any new orders. As such, S. & H. Morton was to cease the building of ships at Leith towards the end of 1911 – and waiting in the wings for the right space and time to take over was another well-known name in Leith shipbuilding.

# THE INTRODUCTION OF HAWTHORNS & CO. INTO SHIPBUILDERS' ROW

In 1911, S. & H. Morton gave up the lease on the land where its shipyard had been in operation for some 60-plus years, to move back completely to its original premises at Bo'ness. Its Leith yard was subsequently taken over by the firm of Hawthorns & Co. which, with a full order book, quickly began building ships.

Hawthorns & Co. had originally been a part of the large Hawthorns of Newcastle Company, which had bought the firm of Maxton's Leith Engine Works and so was involved in the building of steam locomotives. The site was used to build railway engines before a separate new company was formed under the name of Hawthorns & Co. and in around 1881 it set out to build small ships. Interestingly enough, the company under the name of Hawthorns & Co. Leith Engine Works was to go on to build some 400 locomotives from 1846 to 1872. One of these, the Blackie (built 1859, Works Number 162), was the first to be used in South Africa, where she can still be seen at the Cape Town railway station concourse, standing proud as a static exhibit.

The company also had a patent slipway at Granton where it undertook a lot of ship repair work including the lengthening of vessels. It was no small job to lengthen a ship, as this involved bringing the ship onto dry land and then cutting the ship in half, before inserting a suitably designed and built new mid-ship section. This was a job the company had done for some of the Forth ferries originally built by S. & H. Morton.

At the time of its planned move to the S. & H. Morton site, the company still had ships to complete at its old Junction Bridge shipyards, concluding with the launch of the light vessel,

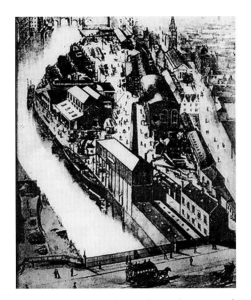

*Looking to the north-east down the Water of Leith towards Leith Harbour and the Firth of Forth. Junction Bridge is in the foreground; Hawthorns & Co. Leith Engine Works, which built railway engines and small ships, is in the centre of the picture. (Author's collection)*

*Alarm*, for the Mersey Docks and Harbour Board. Launched just after her was a sister ship named *Tern*, along with a steam yacht still being constructed. Hawthorns had built more than 135 ships at the old location in the 30 or so years it had been building. While many of these vessels were small steam trawlers, it had also built some large steam yachts that were as good as any built by Ramage & Ferguson.

This all goes to show that Hawthorns & Co. should not be understood as the new guy on the block, busy as the firm had been at its site on the Water of Leith – a site where many of the older shipyards used to be, all launching ships into the narrow river; but there was a restriction on the size of ship that could be built at that location. When the opportunity presented itself to have a shipyard with sea-facing launches, like S. & H. Morton's, of course Hawthorns would take it.

The narrow tidal river, along with the encroachment of buildings on the three other sides of the old yard location at Sheriff Brae/ Coal Hill, close to what would become Leith Hospital, meant that expansion of the shipyard was all but impossible. And with the ever-increasing size of steamships, the company needed larger premises if it were to compete in this growing market. Building up-river on the Water of Leith had its drawbacks in addition to the tidal nature of the river. Ships had to be manoeuvred under two swing bridges, one of which had been the largest in Britain when first built.

Some very fine steam yachts were constructed at the firm's old Junction Bridge yard. The *St Serf*, to be launched in 1913, at 160 feet in length and at 653 TM was a long-life ship, only broken up in 1966 after having many owners. Other examples included the Cox & King-designed *Surf (II)* (not to be confused with the other yachts built by Ramage & Ferguson for the same owner). Hawthorns also built the very beautiful steam yacht *Venetia* (II) in the following year, 1903. She too was still around in the 1960s, both vessels serving as a fine testament to the shipbuilding skills of Hawthorns & Co. The challenge at the Junction Bridge site had been launching ships at an angle to a bend in the river, to be able to get the maximum amount of water available during the launch. This method, of course, also restricted the size of vessel the yard could build and launch. On its new site, however, Hawthorns would be able to launch ships directly into the sea, so the only restrictions now were the length of the building berth and the tide's determining at what time a ship could be launched.

In addition, three steam ketches were built by Hawthorns & Co. and used as deep-sea and hospital mission ships from 1899 to 1902. It also built lightships to go around the British coast, along with lightships for the Irish Light Commissioners and stern trawlers for the fleets around Scotland and further afield – for a Spanish owner, for example. Whalers, too, were within its remit, for C. Salvesen of Leith; the whaler *Scapa* was launched in 1909, a ship typical of its day, at 100 feet long by 19 feet in the beam and 12 feet in depth. These vessels were built for chasing the whale in the cold waters of the South Atlantic.

Aside from the building of ships, the company was very active in repair work and the reboilering of ships, repairs also being undertaken at local collieries and paper mills. It was responsible for the creation of a few large slipways, such as the 2,000-ton-capacity slip for Earnshaw Slipway & Engineering Co. of Manila, and it also built a set of haulage machinery for an 800-ton slipway for Scott & Sons of Bowling, along with further slipways for a firm in Constantinople and another for an Italian company in Naples. Hawthorns' future looked bright.

Incidentally, 1912 also saw much repair work being carried out by Menzies & Co., the famous old company in Leith, which was now no longer building ships. It had further repair work to complete for the Admiralty, having just finished work on overhauls and repairs to a section of the new torpedo destroyers now employed in the Royal Navy. This work was to continue, with further repairs to the second flotilla in what was new work for the Leith yards. Then 1913 would see Menzies & Co. carrying out a record number of dry dock repairs for Leith, with a total of 200 ships being dry docked.

The first ship to be built and launched by Hawthorns & Co. after its takeover of the shipyard of S. & H. Morton was a very large insulated barge of some 1,008 grt, an order for Argentina. She was launched in May 1912, named *Benito Villanueva*, and quickly followed by a small tug for Brazil, part of a two-ship order. The *Aymore* (continuing in the tradition of continuous consecutive order numbers given for the build, she was Ship No 131 on Hawthorns' order book) at 150 grt was pretty standard for a tug built at this time, and she was launched in December 1912, beginning a new line of ships to be built at Leith adjacent to the Ramage & Ferguson yard. The later firm of Henry Robb would also go on to build a few tugs for the Brazilian market – no doubt capitalising on the goodwill and connections already built up, before eventually gaining control of all the remaining yards on this small foreshore.

The next ship was Yard No 134, the *St Fergus*, a passenger/cargo steamer of 390 grt. She was launched on 19 June from the Victoria Shipyard, another order for the North of Scotland, Orkney & Shetland Steam Navigation Co. of Aberdeen, (Previous orders from that company had been won by Ramage & Ferguson, Hawthorns' next-door neighbour.) She was 150 feet in length, with a beam of 24 feet and depth moulded of 11 feet, to be powered by a triple-expansion engine fitted by the builders to produce approximately 450 hp. Her holds had been adapted to carry cattle and other livestock, and she had all the latest machinery for an up-to-date coaster. She was followed in October by the *Lucena*, Yard No 135, for service in Argentina.

Hawthorns & Co. would now begin on the construction of an interesting double order from the Admiralty: two battle-practice targets for the Royal Navy. They were 160 feet in length and 13 feet in depth, heavily constructed to give as much resistance as possible to shot, and designed to float deeply, presenting only around 18 inches of freeboard. On the deck of each low-lying hull would be erected the target. The lower part of the hull consisted of steel packed in timber and cork, and the upper part was Oregon pine.MV *Ila*, Yard No 136, and MV *Ife*, Yard No 137, were a two-ship order from the Elder Dempster Line, both some 300 grt. While they were almost identical, the *Ila*, at 135 feet in length with a beam of 23 feet and a moulded depth of 10ft, was a little shorter and narrower than her sister ship. *Ila* was launched at Leith in January 1914, and then the *Ife*, at 140 feet 6 inches with a beam of 25ft and the same moulded depth as her sister ship, was launched in March. On her sea trials the MV *Ife* attained a constant speed over the measured mile in the Firth of Forth of 8.3 knots, over her specified contract speed of 7.5 knots, so all involved with the vessel were delighted with her performance and that of her new motor engines. The ships were both to be powered by two sets of Bolinder direct-reversible crude oil engines producing approximately 130 bhp. Both ships were twin-screw coasters for the West

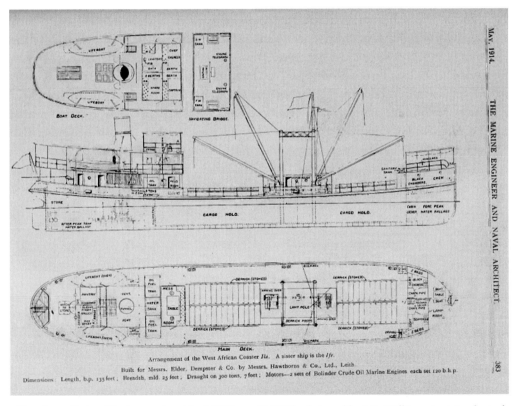

*The above G. A. of the two motor ships MV* Ila *and MV* Ife, *designed by J. B. Wilkie, superintendent of the Elder Dempster Company Liverpool. (Engineering & Shipbuilding magazine)*

Africa service, Lagos being their home port. As soon as the ships' fit-out had been completed, they set out under their own power for West Africa. Both ships were scuttled at Badagry Creek, near Lagos, in 1935. Although they had been built at Hawthorns' Junction Bridge yard they were very significant builds at Leith, and would no doubt have aroused much interest at the Ramage & Ferguson yard, whose new manager, Henry Robb, was now in position.

It is interesting to note that the two Elder Dempster ships were for motor power (diesel) at a time when the marine diesel engine was very much in its infancy, not long after the very first diesel ships were being trialled. Many well-known firms in Sweden and Germany were working on perfecting the engines; the brilliant Herr Junkers was at the forefront of such work in Germany. The Swedish, being the pioneers of this form of marine propulsion, would later license out their very successful diesel engine as the British Polar Engine, to be built and fitted in the U.K. on thousands of ships.

Elder Dempster had been doing some experimental work on barges fitted with diesel engines in service in West Africa, using the well-known Bolinder engine of a smaller size. The next logical step was to see if it could have a successful larger size ship built, using the same type of power source – hence the build of MV Ila and MV Ife, which by all accounts were very successful. The motor-powered ship was now taken seriously as an alternative to steam power, and at least a dozen such ships were built and launched on the Clyde during 1913, and early into 1914 as well.

Hawthorns also won an order for the build of three steam trawlers from a local trawler owner, T.L. Devlin of Newhaven. Each trawler was to be 118 feet in length and intended for the fishing around the Faroe Islands. They would be the largest trawlers registered to the Port of Granton when launched around September and October 1914.

Every ship built supported many outside ancillary and supplier jobs; everything from the supply of cutlery to rope to machine tools, and many local companies benefited from the opportunities to supply the required parts and machinery for the completion of a new build. Similarly, local shops and businesses benefited from the wages earned by the men; most

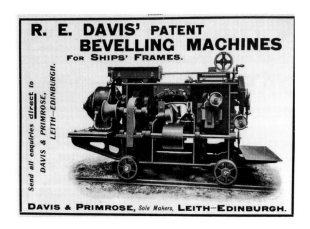

*Davis & Primrose supplied shipyards such as the Royal Dockyards for the frame-bevelling of the latest super-Dreadnoughts being built, along with supplying yards such as Harland & Wolff at Belfast and Govan on the Clyde. (Engineer & Naval Architect magazine ca 1890)*

would spend their wage locally, some even spending it all in the many hostelries around the shipyards.

Many of the local suppliers were very enterprising: steering gear was supplied by Brown Bros. and Davis & Primrose of Bangor Road, Leith, which could offer its own pioneering patented frame-bevelling machines. From around 1866, this latter company, which had started out with premises in Duke Street, made many different machine tools for heavy industry, including steam hammers and steam cranes for shipyards.

Sadly, the firm would go out of business in 1932 along with many more of this type as the huge depression of the early 1930s descended.

Meanwhile, next door to Hawthorns, it was all large steam cargo ships being built – the Ramage & Ferguson yard would not build another luxury steam yacht until well after the end of the First World War.

Following the assassination of Franz Ferdinand in the summer of 1914, it came as no great shock that first Europe and then the world were plunged into war – although the scale of the savagery and the losses would come as a surprise to many. Preparations for war had been ongoing with the Senior Service (Royal Navy) for some time prior to its declaration. Britain had signed the treaty to protect Belgium and on the German incursion into Belgium Britain would keep to the treaty. As Britain was an island, its greatest strength was the sea that surrounded it. So steps were taken to secure the coast, and one of the first changes was the cessation of the excursions in the much-loved and well-used steamers that plied the waters of the Forth. At midnight on 2 August 1914, the Admiralty declared that the Firth of Forth was now a controlled area for military purposes, with excursion sailing prohibited. Britain and the Commonwealth formally entered the First World War on 4 August 1914 with a declaration of war against Germany and the Austria/Hungary alliance (which were, with the Ottoman Empire, to become the Central Powers).

As such, the defence booms would be arranged to restrict the movement of ships and protect the sea lanes of the Forth; they were laid across the Forth from Fidra Island at North Berwick to Elie on the Fife side, with a further boom laid from Granton to Pettycur in Fife. The final boom, with torpedo nets, was laid between Cramond Island and Aberdour, using the islands of Inchmickery and Inchcolm, to help protect the Forth Bridge. Gates were positioned in the booms to allow ships to pass through, and boom defence vessels were stationed there to open and close the gates.

One of the first acts of the war was to seize all German merchant ships berthed in British ports, to which the Germans of course reciprocated, entrapping as many British ships as possible while still in ports controlled by them or their allies.

So began the war to end all wars.

# SIX: THE FIRST WORLD WAR, 1914–1918

As was to be expected, the shipyards of Leith very much played their part during the 'war to end all wars', as the First World War became known at the time. The Victoria Shipyards kept busy, ultimately building a total of 60 ships between them during the four long years of conflict, even though none of the vessels could be classed as warships. Whether this was due to an Admiralty decision or to the size of the yards is mere conjecture at this stage, but more than likely the Admiralty preferred its warships to be built by its own Royal Dockyards. Before, and to a certain extent during, the Second War this opinion persisted in some areas of the Admiralty – commercial yards could not really build warships, or if they did they were not of the same high quality as those built by its own Royal Dockyards. It was perhaps a kind of snobbishness that existed amongst some of the elite who made up the Board of the Admiralty.

That said, some of the larger shipyards of the British Isles *were* entrusted with the build of some of the capital warships, at which point they proved that they were capable of building very fine dreadnoughts and large armed cruisers. It was also a matter of necessity; the Royal Dockyards were fully stretched and the country needed large warships to counter the slightly smaller but still sizeable German fleet of the period between 1914 and 1918. And that was fortunate for Britain, because otherwise not just the First World War but also the Second would probably have had a different outcome.

The yards at Leith did build some vessels for the war effort, including hospital ships and landing craft, but most of their output remained in commercial ships, which were then taken over by the Royal Navy for use as auxiliaries or similar. And of course, a great many Leith-built ships from previous years had been taken over by the Admiralty for use in the various duties required by a navy in time of war.

Inevitably, a great number of these ships of all provenances were sunk or destroyed during these four years of world madness.

Once it was understood, within six months of the outbreak of the First World War, that merchant ships were regarded as legitimate targets by the German U-boats, the British Admiralty encouraged merchant vessels to arm themselves with deck guns, ostensibly for the purposes of defence. Some of these merchant ships, in the great traditions of British seafarers, took it upon themselves to actively attack German shipping, often using false flags. The German Empire, as one might expect, grew to view these vessels as belligerents rather than as neutral shipping, a role they had initially been accorded by international law. Hence, the Germans decreed that ships so engaged were in fact pirates and could blame no one but themselves for falling foul of submarine attack or attack by warships of the Axis powers – and such was war that this meant any ship that was British or an ally of Britain.

(Some of these ships and what happened to them are described in Chapter 7).

Several ships went on to do terrific work during the war at sea, both for the Royal Navy and in the later part of the conflict for the U.S. Navy. Many of them would survive this time and go on to provide great service in the Second World War just 21 years later. Nevertheless, the casualties were immense, and the ships built at Leith suffered as much as any others, so a lot of the work carried out at this time was in the repair of damaged vessels. In 1917, Leith's slipways were extended to facilitate the building of larger ships, along with the hope that once the war was over the Leith yards would get a share of the much-anticipated new builds – thinking shared with every other shipyard in the British Isles.

On the Ramage & Ferguson stocks at the outbreak of the First World War was a very large ship intended as a five-mast sail training ship for the Danish Navy. She was on the stocks for almost two years, and when launched, Yard No 242, she was taken over by the Admiralty as an oil storage hulk and given the name *Black Dragon*. Under the Admiralty she stayed moored in Gibraltar for many years, eventually to become something of an eyesore, tethered to her mooring buoys at this huge navy base on the southern tip of Spain, strategically placed for any shipping movements into or out of the Mediterranean.

She was subsequently renamed *C.600* and was broken up in 1960.

Then in 1921 a new ship was built using the original drawings, and launched as the *København*, the largest sailing ship ever built in a British shipyard. Her magnificent hull, with her long sweeping graceful lines was, then, built twice and who knows, but for the twists of fate that war brings to man, what she might have become if it had been possible for her to be used as a sail training ship, as originally intended? There might have been no mysterious disappearance of the *København* with the terrible loss of life this incurred.

When war broke out in 1914, the Ramage & Ferguson yard was working on the build of a couple of large steam passenger ships for the famous British India Steam Navigation Co. Shipping Line. These were SS *Chakdara* (ON 136317), Yard No 238, and SS *Chakdina*, Yard No 239. Both were launched that year, in June and September respectively, both at more or less the same gross tonnage (just over 3,000 grt). Originally built to carry passengers and cargo, they were sizeable vessels for the Leith yard to build.

*Chakdara* was launched with the following official details released to the maritime press of the day:

The vessel is one of two of the same class fleet, and is intended for her owner's Eastern trade, with dimensions of 330 feet 7 inches in length, with a beam of 46 feet and a depth of 24 feet and 6 inches, with tonnage at some 3,100 tons. She was built of steel under Lloyd's special survey for their highest class and would have a Board of Trade passenger certificate. Powered by one 3-cyl. triple-expansion engine, driving a single shaft, to produce some 3,200 ihp, to give her a speed of some 15 knots. Her accommodation provided midships for first class passengers in two-berth cabins, with second class passengers being accommodated in three-berth cabins, and with first-class saloon and sitting room at the fore end of the bridge and second-class saloon under the aft end of the bridge. Provision was made in the tween decks for carrying a large number of native passengers (1,450 in total deck passengers), her engines will be supplied with steam from four large single-ended boilers working at 215 lbs pressure.

At a build cost of around £84,000, she was the lead ship of the two-ship order (with the *Chakdina*), and she had a further sister ship named *Chakla*, built elsewhere. *Chakdara* was the first of the *Chak-* sisters to be delivered to the British India Line, and also the first ship to be built for the long-established British India Steam Navigation Co. on the east coast of Scotland. She was handed over to her new owners who were delighted she had performed admirably on her sea trials, but towards the end of 1915 she was requisitioned by the Admiralty and used as a troopship for the Indian Expeditionary Force. She was continuously in this service as a troopship for the full duration of the First World War, mainly between India and the Persian Gulf or Suez.

One soldier's report of his trip home from war in 1919 aboard the *Chakdara* mentions how he was not very impressed with the ship. The voyage was from the Persian port of Basra up through the Red Sea to Port Said in Egypt, and was undertaken after the end of the war, in January 1919. The rest of the repatriation journey would be taken up by larger ships returning to Britain from Australia and New Zealand. The soldier was a machine gunner who had been seconded with many others to aid the Indian forces fighting all through the Middle East against the Turkish/Ottoman Empire forces. He complained about both his accommodation on board and the poor food served up. It was not the ship's fault, but he probably had a point. She would have been hard run during the four-year stretch of war, only bare maintenance would have been carried out, and her first- and second-class accommodation would have been taken up by officers or for any other passengers carried at the time.

SS *Chakdara* continued in service with the British India Steam Navigation Co. for a further 14 years of service on the route of Calcutta–Chittagong–Rangoon service before being sold on by the company. She was purchased by the Burma Steam Navigation Co., Bombay, in 1933 and renamed SS *Burmastan*, but stayed with this company for only a few months before being sold on to the Scindia Steam Navigation Co., and retaining the same name. Then SS *Burmastan*, on voyage from Chittagong to Rangoon with general cargo, struck

*HMS* Chakdina, *seen in the above picture sometime in 1941, armed, and complete with her dazzle camouflage paintwork. (Author's collection)*

a submerged rock at Kyaukpyu Harbour, Burma, on 25 July 1935. Although she was refloated the following day, she sank once more and was declared a constructive total loss.

Her sister ship, SS *Chakdina*, was also taken up by the Admiralty for service as a troopship, working continuously with the Indian Expeditionary Force in the Middle East and India. When she was returned to service with the British India Steam Navigation Co. in 1919, it is interesting to note that her deck passenger capacity had risen to a total of some 1,620. Perhaps this was a good enough reason for the company to keep hold of this vessel; she was still with it when the Second World War broke out in 1939. SS *Chakdina*, in the service of the Royal Navy as an armed troopship, was renamed HMS *Chakdina*. She was to go on and serve during the early days of the Second World War, but with tragic consequences for the New Zealand and other Allied soldiers she was carrying to safety out of the desert battlefield that was Tobruk at the time; while offloading troops there, she was bombed and sunk with great loss of life.

The next ship to be completed was SS *Westmoreland*, an order from Donald Currie & Co. for its Liverpool & Hamburg Steamship Co. Ltd. She was a fair-size passenger/cargo vessel of 1,765 grt, with a length of 210 feet and a beam of 31 feet and moulded depth of 16 feet 2 inches. Launched in January 1915, she was powered by a triple-expansion engine built by Ramage & Ferguson, giving her a service speed of 12 knots. She was immediately taken up by the Admiralty for service as a hospital ship/transport ship, and she survived the war. Once taken back by her original owners, she was sold on in 1919 to the Cork Steamship Co. Ltd, which renamed her SS *Cormorant*. She saw service with that company from 1919 to 1933, when she was sold once more to Scotto, Ambrosino, Pugliese & Co. of Oran, French Algeria. She was now a French ship, renamed SS *Oranaise*, and served for a further three years before she disappeared.

Her disappearance has never been satisfactorily explained. It is known she was en route to Sète in southern France on 13 August 1936, but French press reports at the time were full of inconsistencies and contradictions, as was the official enquiry. She had a crew of 25 men and was carrying 15 passengers when she disappeared about 40 nautical miles off the coast of Algeria. Speculation as to her loss ranged from hitting a Spanish mine to being pooped by a huge rogue wave – very unlikely in the Med, so it might just have been bad storage of her cargo of wheat that perhaps shifted or even exploded. There was no SOS sent, so whatever happened, happened very fast. It was said that there were only two survivors from the vessel

– two Djibouti stokers – although how they survived and got off the ship is not recorded. It was further reported that one of them had lost his mind, while the other presented some very incoherent testimony. Of the 38 victims of the sinking, one was the son of the founder of Scotto, Ambrosino, Pugliese & Co., and he was at that time the chairman of that company. Ultimately, SS *Westmoreland* simply joined the long list of ships sunk under mysterious circumstances with large loss of life.

SS *Beltoy*, Yard No 241, ordered for the Shamrock Shipping Co. Ltd of Larne, was the next ship to be launched, in April 1915. She was a standard cargo steamer of 1,544 grt, and the lead ship in a two-ship order. She had a long and varied history, surviving both world wars. She was involved in not one but two collisions within the space of a month in winter 1943, the first being the most serious. This occurred on 16 January in conjunction with SS *Longbird*, which had been travelling independently having left Convoy WN 380 after arriving at Methil, in Fife. On her journey south from Methil *Longbird* collided with SS *Beltoy* some 10 miles west of Tynemouth pier. The damage to the *Longbird* resulted in her sinking, while the *Beltoy* appears to have escaped relatively unscathed prior to her collision the following month, this time much further north in the naval anchorage of West Weddell Sound, Scapa Flow, Orkney. RFA *Attendant*, a coastal oiler, was her adversary this time, on 12 February, and *Attendant* suffered damage but it is not known if the *Beltoy* did likewise – she just carried on in her charmed way, subsequently surviving the Second World War as well. Records show the *Beltoy* was renamed *Min Jem* (Chinese) in 1946, and then again in 1947 to *Hai Nu*, and once more, to *Aquadulce*, in 1949, before being deleted from Lloyd's Register in 1956.

As SS *Beltoy* was on the stocks being constructed, the Admiralty placed orders with many of the smaller shipyards around the country for some 200 lighters to be constructed. Five of the order were to be built by Ramage & Ferguson. In February 1915, orders were placed for the design of a purpose-built landing craft. This design was created in four days, based on the existing Thames Barge design but with the addition of a spoon-shaped bow to take shelving beaches and a drop-down bow ramp. The craft were of a pretty simple design and not too complex to build, which was just as well, as they were required as soon as possible for the invasions being cooked up by the First Lord of the Admiralty, Winston Churchill.

The X-lighters – or, as they were to be nicknamed by the soldiers, beetles – were made to carry 100 men, with a light displacement of around 135 tons. They were 105 feet in length, with a beam of 21 feet and a depth of 7 feet 6 inches, and had armour on the sides to try and stop bullets. Powered by a simple engine using heavy fuel oil, they had a design speed of around 5 knots. The spoon-shaped bow for ease of landing on steep shelving beaches, and the bow ramp, were the only semi-complex parts of the build, as they also had a parallel bottom over some 60 per cent of their length. These crafts were not designed to be at sea for any great length of time. The original plan had called for them to be armed with a 303 Vickers-Maxim gun, but due to a shortage of guns these were not fitted to the first vessels. Some of the craft had fitments to carry horses, while others would carry the men onto the beaches. They were to see action at the lost cause of the Dardanelles campaign, in the battle for Gallipoli. The navy, unable to remove the Turks, dug in around the entrance to the Dardanelles, and the army was called in to try and make a landing around the area of Suvla Bay. In the operation of getting the men from the ships to the shore, the X-lighters performed successfully, and by 6 August 1915 they had landed many of the men of IX Corps. The X-lighters were used continuously in this terrible campaign, and finally, in January 1916, after the loss of some 220,000 Allied casualties, the army was withdrawn and the X-lighters brought out the troops from the hell that was Gallipoli.

After the war, many of the X-lighters were sold off to shipping firms or foreign governments. However, five were used in the Dunkirk evacuations of the Second World War, as ferries to take the troops from the beach to the waiting larger ships. Of the ones that returned, some were still to be seen around the British ports in the 1950s.

X 94 and X 95 were both launched in April 1915. The slightly smaller X 218 and X 219, known as motor store lighters, were 98 feet in length with a smaller light displacement of 120 tons; they followed in December 1916 and January 1917 respectively. Of the five X-lighters built by Ramage & Ferguson, the first, X 93, was renamed Beaker and used as an oil or water tanker. X 94 was sold on in 1920, while X 95 was wrecked in June 1940. X 218 was seen in the U.K. in the 1920s, having been sold out of service for commercial use in 1920, and X 219 was converted into an auxiliary transport, also sometime in the 1920s.

The war was to be brought right to the shores of the shipyard workers when the first Zeppelin air raid on Edinburgh was carried out in 1916. On the nights of 2 and 3 April 1916, the German Zeppelins L14 and L22 dropped something like 23 bombs on Leith and the city of Edinburgh. On the first night, the Germans attacked the British mainland from the air alongside surface ships and submarines in a combined operation. In all, 13 people were killed and 24 injured during the two raids. The chief constable of Leith noted in his official report that the two airships followed a route from Leith to Edinburgh using the Water of Leith to guide them. All the bombs dropped fell within 100 yards of the river, the first few bombs landing in Leith Docks.

The bombs which landed in and around the docks damaged business premises and the quays and a large number of windows in shops, offices and dwelling houses. Two bombs landed in Commercial Street, less than 250 yards from the shipyards. A further explosive bomb fell, according to a police report, 'on the roof of a whisky bond belonging to Innes & Grieve, wholesale spirit merchants, setting fire to the bond, which was entirely destroyed'. The Leith chief constable also reported that some 20 bombs fell on Leith during the second night, killing two people. More bombs were dropped on the south side of Edinburgh, inflicting more casualties and damage to property including the deaths of a further five unfortunate civilians asleep in bed. Following the raids, the same chief constable was to question whether the current air raid precautions were sufficient, and he drew attention to the fact that those precautions only covered the suppression of electric light and not gas light – and moreover a large number of ships were at anchor in Leith Roads, just outside the port, all with their lights burning away. It was reported at the time that it was not until the bombs began to fall that the lights on the ships were extinguished, by which time the Zeppelins had pinpointed the port and by following the glare of the river made their way to Edinburgh to bomb the city as well.

As a result, there followed an Admiralty order that all ships at anchor in Leith Roads and around Granton would have no lights of any description visible outboard. A further outcome from the authorities was that in the event of an air raid warning in future, all lights, be they gas or electric, would go out. With the advent of bombing from the air, the home population were now subject to the constant thought that at any time during the night death could fall upon them from the sky unannounced. This was also true of the shipyard workers; they were now targets if a raid should happen in daytime working hours. It is no surprise, then, that a certain trepidation crept into the normal working day. It was also at that time, with so many men, including shipbuilders, fighting on the fronts in mainland Europe or already dead, that the shipyards turned to women to fill the need for workers.

The war at sea was becoming one of attrition, and British ships were being lost at a rate that could not be replaced by the shipyards in Britain alone. After U-boats had sunk seven US merchant ships, U.S. President Woodrow Wilson went to Congress, calling for a declaration of war on Germany, and this was voted for on 6 April 1917. And so now the might of the American industrial machine was trying to get to grips with the demands of war.

It would take the American shipyards some time before they were able to produce ships in any kind of the numbers required. Take the month of October 1917: some 240 ships were lost by all belligerents, and not all to submarines; some had hit mines, while others were involved in collisions or skirmishes. British merchant ships flying the red ensign – the British merchant flag – accounted for a total of 100 ships lost that month, and then there were the Royal Navy losses as well. There were only two days in October when a British ship was not lost – a staggering statistic.

The U-boats were a huge problem to Allied shipping. The convoy system would reduce some losses to merchant ships, but other means needed to be developed to contain this new underwater menace of war. During the First World War, unrestricted submarine warfare caused a shortage of tankers. The United States ambassador to the United Kingdom, Walter Hines Page, wrote:

> The submarines are sinking freight ships faster than freight ships are being built by the whole world. In this way, too, then the Germans are succeeding. Now if this goes on long enough, the Allies' game is up. For instance, they have lately sunk so many fuel oil ships, that this country may very soon be in a perilous condition — even the Grand Fleet may not have enough fuel.

Against this backdrop, the shipyard of Ramage & Ferguson launched two sister ships. The first, SS *Darino*, 1,433 grt, was launched in July 1917. She was powered by a triple-expansion engine giving her a nominal horse power of some 205. She saw service with the Ellerman Wilson Line of Hull on England's east coast, so that was her home port. She was to stay with Wilson Lines until 1922, when she was transferred to another of the giant Ellerman Group's shipping lines, moving her home port to Liverpool to become an Ellerman

Papayanni Lines ship from 1922, until her sinking in 1939. SS *Darino* (ON 139323) was yet another casualty of war. Although she survived the First World War, she would be sunk during the Second.

The *Darino* was smaller by some way than her sister ship, SS *Serbino*, which was, at 4,080 grt, one of the largest ever built at the shipyard. She was a huge ship for this small shipyard; indeed, she was only built as a result of extending the slipways to accommodate the building of larger vessels in 1917. SS *Serbino* was launched in March 1919 for the Ellerman Papayanni Line's service from Liverpool to the Mediterranean. Powered by a triple-expansion engine giving her around 390 nominal horsepower with a service speed of 10 knots, she was to serve the Papayanni Line well until she also met her fate during the dark days of the Second World War. En route from Mombasa to Liverpool via Freetown, she was sunk by U-82 in 1941, with the loss of 14 of her 65-man crew. Three of these were DEMS gunners, but only four crew members and one DEMS gunner are commemorated on the Tower Hill memorial to merchant sailors who lost their lives in wartime.

It is interesting to note that during 1917 confidence in shipbuilding at Leith was at an all-time high, with plans afoot to extend the existing slipways where possible. Some delays were expected, as Ramage & Ferguson was looking at major excavations of the foreshore, for which the Crown estates had to be involved, as well as the Leith Dock Commission. It would take time, but the yard would get its extended slipways.

Along with this application for planning permission was another, from an as yet un-named company, to carry out the business of shipbuilding at Leith. As of November 1917, it was understood that this scheme would also be looked upon favourably by the Leith Dock Commissioners. It came to light later that this was the brainchild of Henry Robb, who had now put his plans in place to leave Ramage & Ferguson and strike out on his own, taking with him some men hand-picked for the purpose.

Ramage & Ferguson did not launch any ships in 1918; all hands were occupied on repair work. Not only were there so many ships that needed to be repaired urgently and put back to sea, but also there was constant direction to convert any type of merchant ship for war purposes. Conversions were carried out on the larger vessels for use as hospital or stores-carrying ships, and trawlers were in demand to be converted for mine laying or submarine chasing.

Public records from just a month before the end of the war indicate that the huge Ellerman Group had not been slow to act in gaining control of the small firm of Ramage & Ferguson. In October 1918, the share records of that firm showed that John Reeve Ellerman, ship owner of London, had one share while the company secretary of the Ellerman Lines, a Percy Easterbrook held one more. And the Ellerman Lines Group of London held the remaining 498 shares of the firm. As mentioned earlier, Richard Ramage had resigned in 1913, with Henry Robb being brought in as the replacement manager. The other partner and co-owner of the company, John Ferguson, had finally sold his shares to Ellerman in the summer of 1918. Thereafter, he stayed on as a director of the company.

# JOHN CRAN & CO.

When war broke out, the firm of John Cran & Co. were well placed to continue building tugs (for which, as we have seen, they were now very well known). As many as 400 tugs would be used by the Admiralty in the four long years of this bloody conflict, the little workboats carrying out sterling work in the name of King and Country. Their roles were varied, and most – aside from the larger Saint class tugs completed by the yard – were unarmed, probably contributing to their many losses.

At the outbreak of war, the Royal Navy only owned seven fleet tugs, mainly civilian vessels that had been purchased as required and used in normal tug duties at the naval ports around the country. With war declared, the Admiralty put in plans to requisition civilian tugs to meet the increased requirements for vessels to work on patrol, along with the huge increase in the need for minesweepers, anti-submarine vessels and the like. In all, around 100 civilian tugs were requisitioned in this way, yet as the war dragged on past 1914 and into 1915 there was now a huge need for more vessels to be used for salvage and rescue work.

As a result, the Admiralty began placing orders with civilian shipyards. It required a variety of vessels, and the Saint class tugs were the largest of the three different standard designs ordered by the Admiralty. These would be armed, and although they were built to civilian rather than naval specifications, they would also be equipped with the cutting-edge development of radio.

Cran & Co. were to fulfil orders for several of the smaller type of tugs. Having launched the first of two steam fishing trawlers at the end of 1914, the firm swung into action, starting with two sister vessels, both 214 grt: the *Vigoreaux*, Yard No 102, launched in March 1915, and the *Courageaux* in June. These were swiftly followed by *A.M. Stewart*, Yard No 104, launched in December 1915. This large and powerful single-screw tug of some 279 grt was quickly converted to a salvage vessel the following year. She was to survive the war and work for several more years before she foundered off South Cape, Spitzbergen, on 26 April 1933; a sad end for a hard-working ship.

*The tug* Bramley Moore, *launched at Leith in May 1916. She was 214 grt and hired by the Admiralty to serve as a dockyard/general towing vessel before being taken up as a rescue tug for the later part of the war. She would also be used (as was the case with most of the hired tugs) as an expeditionary force tug during part of the war. (Author's collection)*

The next build was the *Gladstone*, Yard No 107, a standard-type tug, launched in April 1916, just one month after her sister ship, the *Bramley Moore*. Both were 214 grt. The *Gladstone* was taken up for service in August 1916 and used as an expeditionary force tug alongside normal towing duties and so, although chartered as a naval tug, she flew the red ensign. She was in hired naval service until early 1919.

The next two tugs built were to be taken up by the Admiralty as hired rescue tugs, the *Alexandra*, Yard No 108, being the first launched, in October 1916. She would see service in her designated role until the beginning of 1920 before being returned to the commercial owners for whom she had originally been intended. Her sister ship, *Alexandra II*, was built and launched the following month, November 1916. She was in service from April 1917 until her release back to civilian duties in November 1919, when she was renamed *William Poulsom*.

Next on the stocks was the tug *Stobo Castle*, Yard No 110. She was 271 grt, with an overall length of 110 feet and a beam of 26 feet and depth of 10 feet moulded, powered by a standard compound steam engine by the builder (as were all the other tugs) with around 900 indicated horsepower to drive her single shaft. She was taken up as a rescue tug, doing sterling work during the remainder of the war before going back into commercial service. This old tug would also survive the next war, and was only broken up in 1963 in Belgium: remarkable longevity and another testament to her builders.

War or not, the firm of Cran & Co. was not slow in business matters. John Cran decided it was a good idea to go into business with the increased financial clout that the local Somerville family could provide. The Somervilles had been large investors in Geo. Gibson & Co. ship owners of Leith, over a period of several years, but with the death of Campbell Gibson only five months after his son had been killed on active service in the ill-fated Dardanelles campaign against the Turks in May 1915, the family interest in the famous shipping line diminished.

Nevertheless, the controlling interest in Geo. Gibson & Co. had passed to Robert Somerville and his two sons, who continued investing in ships and shipping from their base in Leith. On 21 September 1917 the firm of Cran & Somerville Limited came into being, with John Cran and James Somerville as major shareholders, along with two further directors: Robert Somerville, and Alexander Bicket of the Alexander Towing Company, one of the largest towing companies in Britain. Further shares were spread primarily amongst Geo. Gibson & Co., and that company and the Alexander Towing Co. provided funding. With this kind of backing, the future ought to have been secured for the small shipyard.

Meanwhile, the building of much-needed tugs continued with two from an Admiralty order for a standard design, designated HS type. The first was launched in 1917 and named *HS-30*, swiftly followed by the launch of her sister ship, *HS-31*. (Confusingly, the HS numbers were also given to other Admiralty tug classes, such as the West and Poultry.) This HS type was to be a classic British coal-fired single-screw steam tug design, built in quantity to replace losses during the 1914–1918 war. All of them began life in naval service, mainly towing barges laden with supplies such as fuel and munitions, between the U.K. and the French channel ports, as well as along some of the larger French inland waterways.

In wartime service, all HS tugs would have been painted grey, but almost all of those that survived the war were sold to civilian users, All HS type tugs were of 154 grt, although some had twin screws for use on the wider canals in France and later, back home, on the canals of the British Isles.

*Mount Manisty* was an ex-Admiralty *HS-30*, Yard No 111, acquired by the Manchester Ship Canal (MSC) in 1926; the MSC must have been satisfied with the build of this type of tug as she was in service with them for many years, only being broken up in 1961. The MSC was to become one of the major customers of the future Henry Robb Shipyard.

Two more HS tugs, *HS-76* and *HS-77*, were in build (launched in early 1918) when work also began on the much larger and much more time-consuming Saint class rescue tugs. The first of these, the *St Blazey*, was only completed for launch two months after the end of the war in November 1918. The yard had received an Admiralty order for five of the Saint class ships, although three were to be cancelled on the end of the war.

*St Blazey*, Yard No 115, was a rescue tug of some 463 grt, with a length overall of 143 feet and a beam of 29 feet with a moulded depth of 12 feet, and powered by twin engines to give an indicated horsepower of 1,250 ihp, driving the ship at a service speed of around 12.5 knots. She was armed with a single 12-pounder gun and equipped with wireless. She arrived too late to serve in the First World War, although she would in fact survive the next war, and in later life would be used as a fleet tug and for target towing before ending up being used as a target herself, when she was sunk off the coast of Bermuda in 1946.

Her sister ship, the *St Boswells*, Yard No 116, was launched in May 1919. Although built to the same specification and tonnage as the *St Blazey*, she was not so lucky, and only survived in service until the summer of 1920 when, on 16 June, she hit a mine and went down off the island of Terschelling, the Netherlands. The other three Saint class ships in the five-ship order, to have been *St Austell, St Breock and St Budeaux*, were cancelled in December 1918.

With the end of hostilities and the signing of the Armistice in November 1918, all thoughts in the small yard of Cran & Somerville turned to the repair and conversion work on offer. Altering and equipping ships that had been surrendered by Germany and its allies would provide much work for the shipyards. Cran & Somerville were involved in dealing with more than 30 such ships during 1919 alone. Along with all the merchant ships that the Admiralty had called into service, there was more than enough work to be spread around the yards. There were also the high hopes that many new builds would be available to tender for, and with luck and the correct pricing, the future of the shipyard should look good.

# HAWTHORNS & CO. SHIPBUILDERS AND ENGINEERS

Following the build and launch of the two motor vessels for service in Lagos, West Africa, in the first quarter of 1914, the shipyard of Hawthorns & Co. settled into the new surrounds of Shipbuilders' Row at Leith, engaging in competition with Ramage & Ferguson on the right-hand side of its own yard and with the smaller firm of Cran & Somerville to its left.

Building continued with a couple of steam trawlers and a smaller fishing vessel before war was declared.

Within a short period of time it became clear that the weapons used by the enemy included fighting against fishing vessels in an attempt to starve the British Isles into some kind of peace talks. With a huge fishing fleet bringing back fish for consumption by a population that considered it as much a part of the staple diet as potatoes, this was an effective direct action that caused major disruption. Such were the number of British fishing vessels being lost that replacements could not be built fast enough, and in addition there was the demand from the Admiralty to requisition the larger steam trawlers for other uses such as mine laying or submarine chasing or as armed escorts for the inshore convoys that now had to use the designated shipping channels down both coasts of the British Isles. Hawthorns & Co. immediately began constructing many steam trawlers, both for fishing and for requisitioning by the Admiralty, and the yard was also heavily involved in the conversions of such ships for battle against the enemy.

By 1917, the commercial trawler fleet of Britain was a quarter of its pre-war size, the rest being requisitioned by the Admiralty. Trawler losses were high, so fish catches were down and fish prices inflated. To prevent the industry from collapsing, the remaining trawler fleet, along with some smaller fishing vessels, were requisitioned and placed in a Fisher Reserve of the Trawler Section under Admiralty control. The scheme was not, however, extended to sailing trawlers or to steam drifters.

The first ship to be launched after the declaration of war was in December 1914, Yard No 140, *Coadjutor* (aptly named: its meaning is 'one who works in conjunction with another') at 207 grt. She was a sister ship to *Commandant*, which had been launched in September 1914. These were just the start; the building and launching of the steam fishing trawlers would continue throughout the four years of war, trawlers with names such as: FV *Braes O'Hax*, FV *Linn O'Dee*, FV *Moray*, FV *Pifour* and FV *Greg Ness*. Whether purpose-built or requisitioned by the Admiralty, all trawlers for the war effort were operated by the Royal Navy, and regardless of the vessel's origin the ship's registration number was then prefixed by HMT. All vessels purpose-built to Admiralty specifications for Royal Navy use were known as Admiralty Trawlers. But while the trawlers were operated by the Royal Navy, they were crewed predominantly by local fishermen, who knew the waters very well and provided sterling service for very little recognition in doing a job that the country desperately needed. The preferred tactic of the Germans was not to waste ammunition or torpedoes on the fishing vessels, but to round them up and tell the crews to take to the ships' boats before a small raiding party boarded each vessel and scuttled her by opening all the sea-cocks. There is no evidence in British records of any German attacks designed to take lives – they simply wanted to sink the vessels. It worked. From the beginning of hostilities to the Armistice, Britain alone lost some 561 fishing vessels directly due to enemy action. And this does not take into account the naturally dangerous nature of this type of work anyway. Many ships were also lost to mines laid by the enemy – and this was of course indiscriminate and designed to cause loss of life – although some ships, by sheer bad luck, hit mines laid by the British that had, perhaps, broken free from the moorings and drifted into the fishing grounds. It was unusual for gunfire to be used in attacks against fishing vessels, but should one be engaged in anti-mine warfare or happen to be armed, as many of the larger trawlers were, then she became a legitimate target for any type of attack.

During both world wars, large programmes were instigated to purchase trawlers to provide a fast and cheap way of providing capable defence. Naval trawlers were widely used anyway, and commercial trawlers were also suited for many naval requirements because they were very robust ships, well-built, and designed to work heavy trawls in all types of weather. However, when it was clear that the need was even greater than could be met by the purchase of existing trawlers, a programme of designs for new builds was set in motion. The Royal Navy settled on three standard designs, the vessels classes to be known as: the Mersey class, 438 tons standard, with a load of 665 tons; the Castle class design at 360 tons standard and 547 tons loaded; and the smaller Strath class at 311 tons standard and 429 tons loaded. A light armament was supplied, with the Mersey, Castle and Straths all carrying one 12-pounder gun, with some having the addition of 3.5 to 7.5-inch 'bomb throwers', and a few also had a 4-inch gun. The thinking was simple: first, by replacing the trawl with minesweeping gear, a minesweeper was created. Second, trawlers have large clear working decks which are suitable for both the bomb throwers and, in later days, depth charge racks. Thirdly, adding asdic and a 12-pounder or 4-inch gun up front created an effective anti-submarine ship.

The trawlers were all ordered in batches. The first batch, 250, was ordered in November 1916, with a further batch of 150 ordered in 1917, and a final batch of `140 ordered in 1918. The orders were spread all around the smaller commercial shipyards of the British Isles, and 19 of the Strath class of standard units were ordered to be built at Hawthorns yard at Leith.

The first of these was HMT *William Hutchinson*, Admiralty No 3818, delivered on 4 April 1918. She was 215 grt, (311 tons light, and 429 tons loaded) with a length overall of 123 feet and a beam of 22 feet, with a moulded depth of 12 feet. She was powered by a triple-expansion steam engine to produce 430 ihp, giving her a speed of 10.5 knots. Her armament consisted of one 12-pounder gun, she was provided with wireless, and she had a crew of 15 up to 18. She served as a hydrophone vessel before being sold out of service in 1921 and renamed *Rochevellen*. She would later go on to serve in the Second World War.

The next trawler was HMT George Hodges, Admiralty No 3820, delivered just two months later, on 20 June 1918. Built to exactly the same spec as the *William Hutchinson*, she served until just after the war when she was then sold on and renamed *Hood*.

The third of the Strath class trawlers to be built and delivered was HMT Thomas Henrix, standard spec, and she was to serve until 1919 before being loaned to the U.S. Navy until 1921, when she was sold out of naval service to be renamed *Crevette*. She was another of the trawlers still around when the Second World War arrived, and she served in this second conflict.

The next trawler to be built was HMT William Handbury, Admiralty No 3824. She was delivered on 30 October 1918, just a few days before the official end of the war. She was sold into commercial service in 1921 and kept her original name, also going on to serve in the Second World War.

The last of the trawlers to be delivered was HMT Henry Harding, Admiralty No 3822. She was sold into commercial service in 1920 and renamed *Ocean Clipper.*

The rest of the huge trawler order was completed with the build of a further eleven trawlers all constructed in 1919, minus the three cancelled ships. (More on these vessels can be found in Vol. II.) With only three orders for the Strath class cancelled at the yard of Hawthorns & Co., this loss of potential work was not quite as drastic as for other smaller shipyards, where the cancelled orders were as many as 20 ships – a huge effect on what the yard had thought was to have been stable work for some time to come. Cancellations were a double-edged sword: pleasing, as they meant an end to the industrial-scale slaying of the First War; distressing, for their adverse effect on the productivity of many shipyards.

*SS* Felixstowe *on trials after her launch in 1918 from the Leith shipyard of Hawthorns & Co.*

Hawthorns went on to construct a couple of smaller coastal steamers. One, named SS *Felixstowe*, was launched in 1918, an order from the Great Eastern Railway Company, a prototype in which it's possible to see the progression to future vessels of a similar design. She was 215 feet in length with a beam of 33 feet and depth of 16 feet with a gross registered tonnage of 892. This solid coaster-type ship was to serve with the Great Eastern Railway Co. through to 1923 when she was taken over by the London and North Eastern Railway Company. SS *Felixstowe* served through the Second World War until 1942 before being requisitioned by the Admiralty and converted to a wreck dispersal vessel at Deptford. She increased in tonnage from 892 to 1,200, was renamed HMS *Colchester* and was put on duty at Sheerness. She also survived the Second World War and was returned to the nationalised British Railways in 1948 before once more being sold on in 1950 to the Limerick Steamship Company and renamed *Kylemore.* This fine ship was only broken up at Rotterdam in 1957.

With the many trawlers constructed alongside some Admiralty tugs – *Skirbeck* and her sister ship of 211 grt *Freiston*, for example – the yard had performed very well during the dark days of the First World War. Hawthorns had constructed a total of 31 ships, most of which saw service. Countless ships were also repaired and converted for war work.

# seven: leith-built at war

While some ships were built at the three main Leith shipyards during the First World War, a large number of already existing vessels took part in it: ships built during peacetime for normal peacetime maritime trade. Many would succumb to the ravages of war or the sea or both. Of the three main shipyards now occupying what was known as the Victoria Shipyards, ships from S. & H. Morton, Cran & Co. and R. & F. Shipbuilders (which became Ramage & Ferguson) were lost during the conflict. There were, of course, similar losses for Hawthorns & Co. but as that company had built most of its ships on its site at Sheriff Brae on the Water of Leith, those ships were not regarded as having come from the Victoria Shipyards site.

Leith played a vital role in the First World War, and with so many Leith men in the regular army, Royal Navy or merchant service – losing more than 2,200 men – the shipyards started to employ women. It was a hive of industry: deep-sea trawlers were built as minesweepers; Ramage & Ferguson produced hospital ships; and yards in Leith and Granton were put to work maintaining vessels and building merchant ships to replace those sunk by U-boats. The port became a hub for troop transport ships, supply vessels and hospital ships.

While the unwelcome distinction of being the first merchant ship to succumb to enemy submarines was not given to a ship *built* at Leith, the first loss of the war was a ship *registered* at Leith, belonging to Christian Salvesen & Co.

Contemporary papers within the archive of the general shipping and whaling firm Christian Salvesen & Co. – based in Leith until the late 20th century – tell of the company's trials during the First World War. Indeed, diary entries of both Edward Theodore Salvesen (Lord Salvesen) (1857–1942) and his younger brother, Theodore Emil Salvesen (1863–1942), record the loss of Salvesen's vessel *Glitra,* the first British ship to be sunk through enemy action by a submarine in the First World War.

*Glitra* was sunk by a German submarine on 20 October 1914, just off Skudenes, Rogaland, Norway. *She* had started life as the *Saxon Prince* at the Swan Hunter yard on the Tyne at Wallsend, where she had been launched in 1882, and had sailed with the Prince Steamshipping Co. until 1895 before being acquired by Christian Salvesen & Co. and renamed *Glitra*. During

Many British newspapers of the day had articles similar to this one:

### BRITAIN AND THE EMPIRE AT WAR

Tuesday 4 August: Britain protested against German violation of Belgian territory, Belgium invaded early on 4th, Germany declared war on Belgium. British mobilisation ordered Britain at war with Germany from midnight on 4th August 1914.

Meanwhile Germany issued a degree with regard to war at sea:

U-boat Warfare – Warships to be attacked without warning; any commerce warfare to be carried out according to International Law and prize rules i.e. ship to be stopped, boarded and examined, either taken into port by prize crew or passengers were taken on board before the ship sunk.

This, clearly, would be a new type of war, and the Germans' policy continued in principle until February 1915.

a voyage from Grangemouth to Stavanger in Norway carrying coal, iron plate and oil, she was stopped and searched 26 km off Skudenes – just outside neutral Norwegian territorial waters – by the German U-boat *U-17* commanded by Kapitänleutnant Johannes Feldkirchner.

No lives were lost during the incident, however, as Feldkirchner had ordered the crew of the *Glitra* into lifeboats. The German sailors then opened the ship's sea-cocks and scuttled her. After *U-17* left the scene, the torpedo boat *Hai* of the Royal Norwegian Navy took the lifeboats under tow to the Norwegian harbour of Skudeneshavn. The same U-boat, *U-17*, captured and sunk Salvesen's vessel *Ailsa* just north-east of Bell Rock in the North Sea on 17 June 1915.

The day before the declaration of war the *San Wifrido*, at 6,458 grt, was mined and sunk off Cuxhaven at the mouth of the Elbe, and her crew made prisoners. This was followed by the sinking of the *City of Winchester*, 6,606 grt, captured and scuttled by the German light cruiser *Königsberg* on 6 August 1914, some 280 miles from the British port of Aden. The first naval loss also occurred on the same day, when the light cruiser HMS *Amphion* was lost. The Germans had clearly begun the war as they intended to go on, by capturing or sinking some 26 fishing vessels on the first day of the war, 4 August 1914.

Not long after hostilities had broken out, Britain began a naval blockade of Germany. This was intended to stop the German High Seas fleet putting to sea, and also to cut off vital military and civilian supplies. Although the blockade violated accepted international law backed by several international agreements signed over the past two centuries or so, it proved to be pretty effective, the German fleet venturing out only once during the war, to engage in the somewhat inconclusive battle of Jutland. Inconclusive it may have been, but the German fleet never took to the sea again.

Britain mined international waters to prevent any ships from entering entire areas of the world's oceans, causing danger even to the many neutral ships still at sea. On 4 February 1915 Admiral Hugo von Pohl, Commander of the German High Seas Fleet, published a warning in the *Deutscher Reichsanzeiger* (*Imperial German Gazette*), translated thus:

(1) The waters around Great Britain and Ireland, including the whole of the English Channel, are hereby declared to be a War Zone. From February 18 onwards every enemy merchant vessel encountered in this zone will be destroyed, nor will it always be possible to avert the danger thereby threatened to the crew and passengers.

(2) Neutral vessels also will run a risk in the War Zone, because, in view of the hazards of sea warfare and the British authorization of January 31 of the misuse of neutral flags, it may not always be possible to prevent attacks on enemy ships from harming neutral ships.

(3) Navigation to the north of the Shetlands, in the eastern parts of the North Sea and through a zone at least 30 nautical miles wide along the Dutch coast is not exposed to danger.

So began the end of the officer-type class of warfare when gentlemen acted like gentlemen; instead, destruction and killing becoming paramount. With their limited response to the British blockade the Germans naïvely expected a similar response to its unrestricted submarine war against shipping, both naval and commercial, and German U-boats attempted to cut the supply lines between North America and Britain. The nature of submarine warfare meant that attacks often came without warning, giving the crews of the merchant ships little hope of survival. But the United States launched a protest, and Germany changed its rules of engagement.

After the sinking of the passenger ship RMS Lusitania in 1915, Germany promised not to target passenger liners, and Britain armed its merchant ships, thus placing them beyond the protection of the 'cruiser rules' which demanded warning and the movement of crews to 'a place of safety' (a standard that lifeboats did not meet). Finally, in early 1917, Germany adopted a policy of unrestricted submarine warfare, realising that the Americans would eventually enter the war anyhow.

Germany sought to strangle the Allied sea lanes before the United States could transport a large army overseas, but Germany could maintain only five long-range U-boats on station, and to limited effect. Some would contend that the U-boat threat lessened in 1917, yet losses were still very high, some months seeing six or seven ships lost every day. But the British and their allies *were* becoming better equipped to fight U-boats at sea. One of the tactics was not new – the convoy, which had been used years before to foil the attempts of privateers to take ships and goods. Plans were drawn up for the use of convoys escorted by patrol vessels and destroyers, which made it difficult for U-boats to find targets and significantly lessened the merchant losses; and then, after the hydrophone and depth charges were introduced, accompanying sub-hunters and sub-chasers, along with the larger and faster destroyers, could attack a submerged submarine with some hope of success. Troopships, however, were too fast for the submarines and so did not cross the North Atlantic in convoys.

Convoys slowed the flow of supplies, ships having to wait as their convoy was assembled. One of the solutions to the delays was an extensive programme of building new freighters, primarily carried out by the large American shipbuilding industry, but this would all take time and it was only towards the end of the war that ships began to arrive from America in large enough numbers to make a difference.

The U-boats had sunk more than 5,000 Allied ships. Estimates vary, but the final figure could be as high as 7,000 ships lost – not all to submarines, but either way a high wartime cost in men and materials. Estimates of German U-boat losses run to around 199 submarines, showing just how potent this relatively new form of warfare was.

As we have already identified, into this conflict were drawn the ships built or registered at Leith; all kinds of ships to serve the armed forces on land and sea. The following describes some more of the varied services carried out during wartime:

The ferry *Roslin Castle* (ON 123015) was one of many such vessels built at the old Hawthorns & Co. Commercial Street shipyard on the Water of Leith before the company moved down to the Victoria Shipyards, taking S. & H. Morton's place there. *Roslin Castle* had been launched in March 1906 with a length overall of 194 feet and a beam of 26 feet with a moulded depth of 6 feet. At 651 tons, she was powered by a triple-expansion steam engine producing 1,300 ihp to give her a service speed of around 13 knots. The ferry was an order from the Galloway Steam Packet Co., which used her for a couple of years for a ferry service in the Firth of Forth, then sold her to the Admiralty in 1908 to be renamed *Nimble* with pennant number of X76, for use as tender/ferry on the Medway. During the war the *Roslin Castle* worked this route as a special service vessel until 1917, and she was also involved in a prisoner of war exchange service at Boston, Lincolnshire, between 1917 and 1918. She survived the First World War and also saw out the Second before being sold for breaking up in 1949.

The following information about the ex-*Roslin Castle* is from the Royal Fleet Auxiliary Historical Society's website.

### RFA *Nimble*

**Items of historic interest involving this ship:**

Background data: She was originally built as a sloop for her owners' Firth of Forth ferry service. Did not become RFA manned until 1944 when she served on the River Medway Ferry Service, employed on service between Chatham –

Sheerness – ships in the Medway area in connection with the assembly for Operation Overlord

14 March 1905 launched by Hawthorn & Co, Leith as Yard Nr: 110 named *Roslin Castle* for Galloway Saloon Steam Packet Co., Leith

May 1905 completed at a cost of £12,210

8 June 1906, when between the Bass Rock and North Berwick a passenger was seen to jump over the side. He was rescued by a boat lowered from the *Roslin Castle*

23 June 1906 sailed Leith and when off Seafield ran aground on a sandbank. Managed to get off returning to Leith Docks.

7 September 1906 the Sanitory Congress which was holding a meeting at the Queen's Hotel, Leith adjourned to the *Roslin Castle* for a cruise on the Forth from the West Pier, Leith at the invitation of the Corporation of Leith

25 March 1908 purchased by the Admiralty for £15,500 for service as a tender based at Sheerness and was renamed *Nimble*

16 October 1914 arrived at Dover

1914–1917 on the Chatham–Sheerness ferry service

1917–1918 served on PoW exchange duties at Boston, Lincs

21 February 1919 at Chatham alongside HMS *Bacchante* discharging kit bags and hammocks of active service ratings to depot

14 July 1919 came alongside HMS Erebus which was anchored off Sheerness Dockyard8 December 1920 at Sheerness berthed alongside HMS *Yarmouth* to take crew to Chatham with the warship being reduced to Nore Reserve

1922 laid up at Chatham

13 May 1923 at the Great Nore alongside HMS *Repulse* ferrying part of the crew into Sheerness

26 July 1924 used a VIP transport for the Kings Fleet Review at Spithead – the reviewed included RFA *Petronel*

1930–1941 ferry steamer based at Chatham Dockyard on river and harbour passenger service

17 December 1941 laid up at Chatham

2 April 1944 became RFA manned and renamed RFA *Nimble*. Acted as the Medway Ferry Service between Chatham and Sheerness

8 December 1947 placed on care and maintenance basis

11 October 1948 disposal arranged by the Director of Ships and Transport, and was purchased by Lloyds Albert Yacht and Motor Packet Service, Southampton, renamed Nimble

14 February 1949 arrived Boom for breaking up

**Note:** Along with the ferry *Harlequin* was supposed to have been replaced by the purpose-built *Magician* but the latter was turned over to the War Department on completion for use as an army hospital ship, and was not returned to the R.N. until 1945.

Another interesting ship was SS *Vala*, sister ship to SS *Vana*. Built by R. & F. Shipbuilders as Yard No 122, she was launched December 1894 for the firm of J.T. Salvesen & Co, of Grangemouth, Scotland.

She was taken up by the Admiralty and turned into what was known as a Q ship.

This meant that SS *Vala* was armed, but her guns were hidden or disguised. As Germany had declared open warfare on all ships, naval or merchant, this was thought of as a potential way of luring U-boats to the surface – at which point the Q ship would lower the covers hiding the guns and with luck blast the U-boat out of the water.

But SS *Vala* was herself sunk by UB-54 120 miles off the cost of the Isles of Scilly, with the loss of 43 men on 20 August 1917.

The many tugs built by John Cran & Co. played a large part in the war because the Admiralty had very few fleet tugs and even fewer tugs to carry out the myriad duties required. Prior to establishing its shipbuilding programme, it began requisitioning ships, and well over 450 tugs were taken up during the First World War. Some were used by the Admiralty only for a short while and but others were requisitioned for the duration, leading to acute shortages of tugs in many of the busiest port areas of the country. The Admiralty, then, also looked overseas for the vital vessels, and some 50 tugs were hired from countries such as Portugal and the U.S.A. The tugs were put to good use, from home waters, to the coast of mainland Europe, to the campaign at Gallipoli, to the Middle East. They were also used off the coast of West Africa, up to the White Sea off the Arctic, the Black Sea, and in Russian waters.

Some of John Cran & Co's tugs taken up for service included the wonderfully named steam tug *Herculaneum*, Yard No 68, launched in 1909, of 172 grt. She was first a hired patrol tug and then an unarmed patrol ship in and around the Liverpool area. On later service, she worked alongside another tug built by John Cran & Co., the *Hornby*, built in 1909.

The *Egerton*, Yard No 80, was another screw tug hired by the Admiralty. She had been built in 1911, at 272 grt, and saw service from the start of the war until May 1917, when she was replaced by one of the 250 new standard tugs from the first Admiralty new-build programme.

Most of the screw tugs over 70 grt that the Admiralty had hired early on were used as expeditionary force tugs, and replaced by the new purpose-built tugs between 1917 and 1918. Nearly all were chartered as naval tugs and flew the red ensign. Some of the larger vessels were converted for other use, such as the tug *Gladstone*, Yard No 96, built by John Cran & Co. in 1913. At 211 grt she was renamed *Salvage Steamer No 2*, and as her name suggests she was used in a much-needed salvage role.

It was inevitable that many ships would be lost. The aforementioned *Alexandra*, Yard No 108 built in 1907, was another of the hired tugs working in and around the busy port of Liverpool, before being wrecked in October 1915. A further two fishing vessels that had been built in the John Cran yard foundered during the First World War, although not to enemy action – as if to emphasise that seafaring *per se* was dangerous, war or no war. Ships built in the pre-war S. & H. Morton shipyard that were lost to enemy action included the iron steamship *Midlothian*, Yard No 20, launched in November 1871. She was stopped by UB-73 in 1917 while on voyage from Famagusta with a cargo of firewood, about 80 miles south of Cape Greco, Cyprus; her crew were told to take to the ship's boats, but the submarine opened fire on her. The main gun fired a few shells into her before sinking her.

Another old iron steamship to meet her fate during the First World War was the iron steamship *Embla*, Yard No 34, launched in December 1882. She sank in 1915 on a voyage from London to Dunkirk, when on 24 December she hit a mine in the mouth of the Thames; she sank but no lives were lost. The steam cargo ship *Pizarro*, Yard No 38 – by the time of her sinking she was renamed *Punta Teno* – was sailing from Tenerife to Bordeaux with a cargo of bananas and onions when she was sunk by a German submarine in January 1915, off the north-west coast of Spain, with no loss of life.

The steam trawler originally named *Kestrel*, Yard No 41, was sold to a Spanish fishing concern and sunk by shellfire from a U-boat. The steam cargo vessel *Orta*, Yard No 58, owned and operated by Christian Salvesen and launched in 1890. She had been sold on to a French shipping company in 1913 and renamed *Cap Mazagan*. On a voyage from Port Talbot with a cargo of coal, she was stopped by UB-38 off the Longships lighthouse at the far end of Cornwall. The crew were instructed to take to the ship's boats as the ship was scuttled, and there was no loss of life.

Many more ships and men who sailed on the local shipping companies' ships were also lost at sea during the First World War, the Leith, Hull and Hamburg Steam Packet Company Ltd losing more than 20 of the total ships owned, at least 18 of them having been lost to enemy submarines or to mines. With some 36 vessels under ownership at the start of the war, the company was left with 21 ships under its control by the end of hostilities in November 1918.

The total number of ships lost from the Leith yards amounted to half of the total registered tonnage for the port. In all, some 55 ships, owned by only eight local shipping companies, were lost to enemy action. Even though some of the local Leith mariners had had the misfortune to be torpedoed two or even three times, they still signed on for service. Although the total number of vessels lost during the First World War is far from an exhaustive list, it does go some way to show that this port contributed significantly to the war effort, especially in proportion to its resources.

The shipyard of Ramage & Ferguson, having built by far the largest number of ships in the Leith yards, turned out to be the biggest loser. The following is by no means a complete list but it demonstrates the many ships lost to both enemy action and the ravages of the sea from this one relatively small shipyard:

The *Eaglescliffe* had been built on spec by the shipbuilders of R. & F. and launched in January 1881, known then as *Kama* and going through many owners and changes of name. She was an iron steam cargo ship of 871 tons, who met her fate in September 1916 while sailing under the Norwegian flag, as SS *Dania*. She was on a voyage from Onega to Leith with a cargo of timber when she was stopped by U-43 on 26 September 1914, at a time when the U-boat captains still abided by a kind of gentleman's agreement to consider the fate of the ship's crew's along with their preference to conserve ammunition and torpedoes. So *Dania's* crew were ordered into the ship's boats around 7 miles north-east of Nordkyn, and the Germans proceeded to blow her up and sink her with no loss of life. The large three-masted full-rigged ship *Mount Carmel*, at some 1,686 tons a large iron-hulled vessel, was lost without trace, last seen off Pensacola Florida in July 1916. With nothing more ever heard of her, she was believed to have foundered in a storm rather than to have succumbed to enemy attack.

The steamer *Mascotte*, was, however, sunk by the enemy. On 3 September 1916, on a voyage from Rotterdam to Leith with general cargo, she hit a mine laid by UC-6 off the coast of Southwold, and sank with the loss of one crew member. At the time of her sinking she was owned by a Dutch shipping firm which had purchased her from Geo. Gibson & Co. Mines were a constant danger, and very difficult to predict, as they would be laid at night by submarines, so there was no real way of knowing where they had been released. They were often difficult to see, some of them floating just below the surface, waiting to be hit. Mines took a heavy toll.

The steamer *Ancona* was a Jas Currie & Co. ship, Yard No 85 when she was launched in June 1888. She had a varied history, including a maritime disaster when in June 1905 she collided with the Danish training ship *Georg Stage*, which was almost cut in half. The collision happened around midnight when *Ancona* was on her way from Alloa to Konigsberg with a full load of coal, and *Georg Stage* stood no chance of survival. She sank within two minutes and 22 cadets drowned, though the *Ancona* did manage to pick up 58 of the remaining cadets and officers. Amazingly the *Georg Stage* was recovered from the seabed and repaired, and she put to sea again until 1934. *Ancona*, meanwhile, encountered a German submarine in September 1916 and although damaged by gunfire she was not sunk. Her good fortune was, however, only to last until the following year, 28 May 1917, when on voyage with general cargo from Falmouth to Lisbon she encountered the German submarine UC-70. She was sunk 110 miles west south-west of Ushant with no survivors from her crew of 25, and a lone passenger was also killed. *Ancona's* sister ship SS *Ravenna*, another Leith, Hull & Hamburg ship managed by Currie, was another casualty of the war; she was to go down after a collision in 1917 with five crew lost.

The steamer *Moscow* ended her days in 1918 at the hands of the Bolsheviks who scuttled her at Petrograd to prevent her being used any more by the opposing Russian forces. She had been built and launched in March 1891 for William Thomson & Co. (later the Ben Line), and requisitioned by the Russians. She was not the only ship to end her days in Russia. SS *Bintang*, Yard No 177, built and launched in 1901 for the East Asiatic Co., also found herself on the other side, being operated by the Czarist forces. She too was scuttled in 1918 to avoid her being of use to the Bolsheviks, who had risen up against the Czar in 1917.

SS *Minorca*, Yard No 116, was another Leith, Hull & Hamburg ship managed by Currie and sunk by a German submarine in 1917. Being defensively armed, she was torpedoed without warning on 11 December while en route from Genoa to the Spanish port of Cartagena. In ballast, she would have been high in the water, presenting a very easy target. Fifteen lives were lost including the master of the ship when the U-64 fired her torpedoes a couple of miles from Cabo de las Huertas, 100 miles short of Cartagena.

Some of the ships flying the red ensign and built at Leith met with a different fate, ending up under the German flag, having been taken captive while in a German-controlled port at the outbreak of war and claimed as prizes. One such ship was SS *Vienna*, yet another of the Leith, Hull & Hamburg ships, identified at the very beginning of hostilities as being just the type of ship that the German navy could use. She was taken as a prize of war in the port of Hamburg and converted into a minelayer/raider. Renamed *Meteor*, she was only in operation for a few months but still accounted for some 15,000 tons of shipping sunk – most of the vessels flying the neutral flags of the Scandinavian countries. On the first of her missions on 29 May 1915, she was to lay mines in the White Sea area and attack Allied merchant ships carrying much-needed supplies to Russia. She was successful in attacking and sinking three unarmed neutral ships on the surface with gunfire, and her minelaying exploits accounted for another three ships; she then returned to port unharmed in June 1915.

The *Vienna's* second mission took her to the north of Scotland to lay mines in the Moray Firth area. Here, she was intercepted by ABV Ramsey, and when ordered to stop, SMS *Meteor's* German captain, von Knorr was able to manoeuvre her into a strategic firing position against the unsuspecting *Ramsey* – after all *Meteor* was a British-built ship. As the *Meteor* opened fire with all guns, she quickly overwhelmed the *Ramsey*, sinking her quickly. The *Meteor* managed to lift the crew from the stricken vessel and place them under guard as prisoners of war. But word had been passed by the *Ramsey* to a light cruiser squadron operating in the area, and it was soon on the scene, ready to blow the *Meteor* out of the water. With the British ships closing fast, and perhaps not willing to risk the lives of her crew in a hopeless situation – or, for that matter, the British captives on board – von Knorr ordered her scuttled. So ended the life of SMS Meteor ex-SS *Vienna*. (See Chapter 5 for more on the *Vienna*, Ship 188).

The largest ship built at Leith to fall victim of the war was the 4,112 grt SS *Cathay*. Yard No 156, which had been launched from the Ramage & Ferguson yard in 1898. She was an order from Anderson & Co. She was flying the flag of Denmark on a voyage from Aarhus to Newcastle in ballast, when on 5 May 1915 she hit a mine. All 35 crew and 7 passengers were rescued by PTB Hamilton, and landed safely at Ramsgate.

With the cessation of German submarine attacks on merchant shipping for a short while during 1916, some of the Allied merchant ships got a small respite. Nevertheless, many were still lost to attacks, as the cessation was primarily applicable to unarmed passenger ships; defensively armed merchantmen were still regarded as legitimate targets.

The Leith-built SS *Bernicia*, Yard No 232, was one such target, stopped by UB-38 while voyaging in ballast from Rouen to London. Her crew was ordered to take to the ship's boats, then the Germans planted explosive charges on board and opened all her sea-cocks. There were no casualties as she sank some 20 miles from Beachy Head on 13 November 1916.

The next year, the Trinity House vessel THV Alert, Yard No 228, launched in December 1911, was on a voyage from London to Dover when on 15 April 1917 she hit a mine in the Dover Straits. Unfortunately the sinking of this fine ship resulted in the loss of 11 of her crew members. Five of the *Alert's* dead were firemen who were below, stoking the fires that kept the boilers going; they, of all crew members, would have had little chance of escape when a vessel hit a mine.

The ex-SS *Narova*, Yard No 126, while owned by Bland Lines of Gibraltar and renamed *Gibel Yedid*, faced a German submarine while on voyage from Montreal to Gibraltar on 13 July 1917 with general cargo. She was targeted by U-48 and went down 150 miles off the French coast. All her crew managed to take to the ship's boats before she sank.

As the toll of ships lost continued, SS *Farraline* was another victim. She had been built in June 1903 for the London & Edinburgh Shipping Line as Yard No 191. Defensively armed, SS *Farraline* was torpedoed without warning while sailing from Bordeaux to Cardiff with a cargo of pit props. On 2 November 1917 she was sunk by UC-69, 15 miles from the French coast, off Ushant, with the loss of one crew member.

As we have seen, with the German submarine attacks on merchant and passenger shipping relenting somewhat by the beginning of 1916, the main thrust was directed at Royal Navy ships and Allied warships. However, as the war dragged on, the German High Command started looking at any measures which might generate victory. To this end, on 22 December 1916 Admiral von Holtzendorff composed a plan that was to become a pivotal document for the resumption of Germany's unrestricted U-boat warfare throughout the rest of the war. Holtzendorff's plan was based on information contained in a study carried out by Dr Richard Fuss, who had ascertained that if the Germans could sink enough merchant shipping Britain would be unable to build enough ships to continue the war and would therefore be forced to sue for peace within six months.

Holtzendorff proposed sinking 600,000 tons of shipping per month, putting Britain into such a bad position that it would be forced out of the war before the disorganised Americans could intervene. Holtzendorff went as far as assuring the Kaiser that he would give His Majesty his word as an officer that not one American would land on the mainland of Europe. On 9 January 1917, the Kaiser met with his chancellor, Bethmann-Hollweg, and the military leaders at the Schloss Pless (in what is now southern Poland) to discuss measures to resolve Germany's grim war situation. The military campaign in France was bogged down, and with the Allied divisions outnumbering the German divisions by some 190 to 150, the very real possibility of defeat was staring the German army in the face. With the German navy bottled up in its home ports and the British blockade causing serious scarcity of food and materials for the entire German population, the military staff urged the Kaiser to unleash the submarine fleet on any shipping travelling to Britain. Hindenburg, however, advised the Kaiser that the war must be brought to an end by whatever means and as soon as possible.

On 31 January, the Kaiser signed the order for unrestricted submarine warfare to resume with effect the next day, 1 February 1917, leading Bethmann-Hollweg, who had opposed the decision, to assert that Germany was finished. Just a few days earlier, on 27 January, the British commander-in-chief of the Grand Fleet, Admiral Beatty, had observed: 'The real crux lies in whether we blockade the enemy to his knees or whether he does the same to us.'

German efficiency in new-build construction ensured that despite losses at least 120 submarines would be available for the rest of 1917, and the new policy of unrestricted submarine warfare began very successfully for the Germans. In January 1917, prior to the campaign, Britain lost 49 ships; in February, after the campaign opened, 105; and in March, 147; that last figure represented the sinking of a full 25 per cent of all British-bound shipping.

So the campaign was initially a great success, nearly 500,000 tons of shipping being sunk in each of February and March, and 860,000 tons in April, when Britain's supplies of wheat shrank to six weeks' worth. In May, British losses exceeded 600,000 tons, and in June 700,000. In contrast, Germany had lost only nine submarines in the first three months of the campaign. On 3 February, in an appalled response to the campaign, President Wilson severed all diplomatic relations with Germany, and the U.S. Congress declared war on 6 April.

At first, the Admiralty had failed to respond effectively to the German offensive. Despite the proven success of troop convoys earlier in the war, the Channel convoys between England and France, and the Dutch, French, and Scandinavian convoys in the North Sea, the Admiralty initially refused to consider widespread convoying or escorting. As mentioned above, convoys imposed severe delays on shipping, and so it was believed that the convoy system would amount to a loss of carrying capacity greater than the loss inflicted by the U-boats. The convoy system was disliked by both merchant and naval captains, and derided as a defensive measure. It was not until 27 April 1917 that the Admiralty endorsed the convoy system, the first one sailing from Gibraltar on 10 May. As more and more merchantmen from Allied countries were sunk, Brazilian ships took over some of the routes that had been vacated. However, this led Brazilian vessels into waters patrolled by U-boats and when meeting Germany's policy of unrestricted submarine warfare Brazilian ships were soon lost. This drove Brazil closer to declaring war on the Central Powers, and it then did so at the same time as the United States. In May and June a regular system of transatlantic convoys was established, and after July the monthly losses never exceeded 500,000 tons, although they remained above 300,000 tons for the remainder of 1917.

So convoying was an immediate success despite the reservations and resentment of the system by many, both in the Royal Navy and in the merchant fleets. On whatever route it was introduced, it resulted in a drop in merchant shipping losses, the U-boats seeking easier prey elsewhere. It also brought the warships escorting the convoys into contact with the attacking U-boats, leading to an increase in the U-boats destroyed. Now German submarine losses were between five and ten each month, and they soon realised the need to increase production, even at the expense of building surface warships. However, production was delayed by shortages in labour and materials.

So 1917 was a bad year for Allied shipping losses, and for the ships built at Leith too, as they continued to become victims of the war. Another ship to be lost was the Ellerman Lines ship SS *Estrellano*, which had been built at Ramage & Ferguson as Yard No 223 and launched in September 1910. She was on voyage from Oporto to London with a general cargo when she was sighted by UC-71, on 31 October 1917. The U-boat fired her torpedoes and she sank with the loss of three men, 14 miles from the coast of the Île du Pilier, France. Another ship which fell to the damned torpedoes was the passenger ferry *St Margaret*, Yard No 235, launched at the Ramage & Ferguson Shipyard in April 1917.

The year of 1918 was not much better for the Leith-built ships. The very first month saw the Leith, Hull & Hamburg ship SS *Alster*, managed by Jas Currie & Co., sunk by submarine action. She had originally been launched at Leith as Yard No 217 in March 1909. She was a defensively armed merchant ship, on passage from Bergen to Hull when, on 14 January 1918 with a cargo of dried fish, she was torpedoed without warning by UB-62. She went down about five miles from Noss Head in Shetland, but fortunately all her crew managed to get into the boats and make it to safety.

The following month saw the loss of yet another Ramage & Ferguson ship, the second of the ships that made up the fleet of passenger ferries of the North of Scotland and Orkney and Shetland Steam Navigation Company. She was the *St Magnus*, Yard No 229, launched in March 1912:

> At 12.15 hours on 12th February 1918, the second *St. Magnus*, 957 grt, built in 1912, was sunk by an explosion three miles north-north-east of Peterhead while on passage from Lerwick and Kirkwall to Aberdeen and Leith. Captain John Mackenzie was in command.

The *St Magnus* was a defensively armed steamer carrying general cargo, mail and 30 passengers; she had a crew of 29 and was torpedoed without warning, with the loss of five lives, by the German submarine UC-58, commanded by Karl Vesper.

A report in the *Shetland News* 20 September 1917 stated:

> It should be mentioned that, after torpedoing the steamer, the crew of the submarine laughed at the men in the boats, and looked for a minute or two at those struggling in the water, and then submerged without offering the slightest assistance.

SS *Elba*, Yard No 162, was launched on 24 May 1889. She was 225 feet in length with a beam of 33 feet 3 inches, with a moulded depth of 16 feet 11½ inches. Her engines were of the triple-expansion type, with cylinders 18½ inches, 30 inches and 49 inches diameter by a 33-inch stroke. Her large single-ended boiler provided the steam, for her engines working up to 165 lbs pressure, fitted by the builder. SS *Elba* was, in fact, the twelfth steamer to be built for James Currie & Co by Ramage & Ferguson Ltd. As a defensively armed merchant ship, she was on voyage from Cardiff to St Malo, France, on 28 April 1918 in a small convoy of five vessels when she was hit and sunk, without warning, by a torpedo from the German submarine UB-103, commanded by Kapitänleutnant Paul Hundius. Six miles north-west by north off Pendeen, West Cornwall, she sank with the loss of ten men killed or drowned. The UB-103 was to suffer her own fate on 14 August 1918 when she was mined off the Flanders coast with the loss of all 37 of her crew on 14 August 1918.

By mid-1918, U-boat losses had reached unacceptable levels, and the morale of their crews had drastically deteriorated; by the autumn it became clear that the Central Powers could not win the war. The Allies insisted that an essential precondition of any armistice was for Germany to surrender all her submarines, and on 24 October 1918 all German

Some truly staggering figures show the results of the resumption of unrestricted warfare by German submarines on the mercantile fleets.

Unrestricted submarine warfare was resumed in February 1917, and the heaviest losses were suffered in April 1917 when a record 881,027 tons were sunk by the U-boats. 150,000 tons of British shipping had been lost in January 1917, and then 300,000 tons in February; Allied and neutral losses increased in a similar proportion. In April 525,000 tons of British shipping was lost. and then Britain began the full-scale use of convoys in September 1917. In October 270,000 tons were lost, and in December 170,000 tons.

A total 12,850,815 gross tons of shipping were lost to U-boats during World War I; nearly 5,000 merchant ships, with the loss of 15,000 Allied sailors' lives.

Sir Joseph Maclay, the Minister of Shipping, approved four standard designs of merchant ship and placed orders for over 1,000,000 tons of shipping. Britain launched 495,000 tons of shipping in the first half of 1917, but 850,000 tons were sunk in the first quarter alone; by 1918 no less than 3,000,000 tons a year were being launched.

(*Adapted from https://en.wikipedia.org/
wiki/U-boat_Campaign_(World_War_I)*)

U-boats were ordered to cease offensive operations and return to their home ports. The last significant role played by U-boats in the First World War was the suppression of the German naval mutiny that same month, when they stood ready to fire without warning on any vessel flying the red (communist) flag. The Allies stipulated that all seaworthy submarines were to be surrendered to them. Many were subsequently brought into the Firth of Forth and other naval areas to be examined by Admiralty experts, and all were broken up for scrap.

In the time frame that this book covers, some seismic changes occurred in shipbuilding; it had gone from wood to iron and then steel as the primary material of construction; it had gone from ships powered by sail to steam and then diesel. But shipbuilding still had its traditions stretching way back into the past; this would not change. As the guns fell silent in 1918 the world was hoping for a better future. Volume II in this series will continue the story of the ships built at Leith from 1918 to 1939.

# THE MERCHANT NAVY MEMORIAL, THE SHORE, LEITH

The Merchant Navy Memorial Trust chose Leith for a new memorial since it had been Scotland's premier port for more than 300 years and had served as Edinburgh's trading port for more than 700 years.

Many of the beautiful luxury steam yachts built at Leith were pressed into service, and sometimes lost, throughout the two world wars. Full details of the fine service these ships gave can be found in the book I am writing about steam yachts built at Leith.

# SHIPS BUILT AT R. & F. SHIPBUILDERS

This is the complete list of ships built at what was first known as the R. & F. Shipyard, later Ramage & Ferguson Shipbuilders & Engineers Ltd.

The information is presented as in the official shipbuilder's yard list. While this volume only covers the time from the start of the yard to the end of the First World War, the list is of all the ships built at the yard from start to Robb's purchase of the yard in 1934.

*The memorial to the merchant fleet at the Shore, Leith. (Author's collection)*

The list is compiled in order of yard number allocated, not by launch date. Hence, you will see that some vessels high on the list were not, in fact, launched until much later, due to priority given to ships needed first, or in order of preferred customer, or delays with a ship's engines and so on.

Tonnage is given as gross registered tons except for steam yachts, which are shown with an asterisk to denote that the measurement used is Thames Yacht Measurement, as tradition demanded at the time.

| Yard No | Name | Tonnage | Type | Launched |
|---------|------|---------|------|----------|
| 1 | SS SHAMROCK | 591 | Iron hulled ship | Feb 1878 |
| 2 | SY RANEE | | Steam Yacht iron hull | Jan 1878 |
| 3 | SS ALBATROSS | 366 | | April 1878 |
| 4 | SS ZYPHER | 350 | | May 1878 |
| 5 | STORNOWAY | 121 | Iron hull fishing trawler | May 1878 |
| 6 | SY MALLARD | | Steam Yacht iron hull | June 1878 |
| 7 | Barge | | | |
| 8 | Barge | | | |
| 9 | Barge | | | |
| 10 | Punt | | | |
| 11 | Tres Hermanies | | Steam Launch | Oct 1878 |
| 12 | DREDGER | 501 | | Feb 1879 |
| 13 | SY TITANIA | 186* | Iron single screw Schooner rigged Yacht | May 1879 |
| 14 | SS AEOLUS | 497 | cargo ship | Nov 1879 |
| 15 | SS GEORGE GOWLAND | 593 | iron cargo ship | Jan 1880 |
| 16 | SY VANADIS | 207* | Steam Yacht | Mar 1880 |
| 17 | SY ALINE | 229* | Steam Yacht | Apr 1880 |
| 18 | SS WOODSTOCK | 569 | iron cargo ship | May 1880 |
| 19 | V | - | Steam Lighter | Jul 1880 |
| 20 | SY FAIR GERALDINE | 188* | Steam Yacht | June 1880 |
| 21 | SS STARLEYHALL | 610 | iron cargo ship | Aug 1880 |
| 22 | Punt | | | July 1880 |
| 23 | SS EAGLESCLIFFE | 871 | Steam cargo ship | Jan 1881 |
| 24 | SS CRAIGROWNIE | 879 | Steam cargo ship | Sept 1880 |
| 25 | SS WYCLIFFE | 1047 | Steam cargo ship | Dec 1880 |
| 26 | SS PENANG | 623 | iron passenger/cargo | Jan 1881 |
| 27 | SS RANEE | 617 | iron passenger/cargo | Mar 1881 |
| 28 | SY IOLANTHE | 412* | Steam Yacht | Apr 1881 |
| 29 | SS CRAIGALLION | 1028 | Iron cargo ship | May 1881 |
| 30 | SS MIDLOTHIAN | 920 | Ferry Ro-Ro Paddle steamer | Aug 1881 |
| 31 | SS ARDANGORM | 1662 | Iron cargo ship | Oct 1881 |

| Yard No | Name | Tonnage | Type | Launched |
|---|---|---|---|---|
| 32 | HIGHLAND CHIEF | 942 | Iron barque sailing ship | Nov 1881 |
| 33 | SY ROVER | 424* | Steam Yacht | Jan 1882 |
| 34 | SY CANDACE | 268 | Steam Yacht | Sept 1881 |
| 35 | SS CLAN MACKENZIE | 2987 | Steam cargo ship | Apr 1882 |
| 36 | SS CLAN MACGREGOR | 3007 | Steam cargo ship | Aug 1882 |
| 37 | SS CRAIGTON | 2090 | Steam cargo ship | Sept 1882 |
| 38 | SY GITANA | 344* | Steam Yacht | Apr 1882 |
| 39 | SS LOCH ARD | 1604 | Steam cargo ship | Dec 1882 |
| 40 | SY ELFRIDA | 109* | Steam Yacht | June 1882 |
| 41 | HIGHLAND GLEN | 1032 | Aux Sailing ship | Nov 1882 |
| 42 | SY STAR OF THE SEA | 252* | Steam Yacht | Oct 1882 |
| 43 | SS MOUNT CARMEL | 1686 | Iron Sailing Ship | Feb 1883 |
| 44 | SS ECOSSAISE | 849 | Steam cargo ship | Mar 1883 |
| 45 | SS GRANADA | 958 | Steam cargo ship | Apr 1883 |
| 46 | EARL OF SHAFTESBURY | 2079 | 4 masted iron fully rigged ship | July 1883 |
| 47 | SS CARRIEDO | 1082 | Iron cargo ship | Aug 1883 |
| 48 | SY MERRIE ENGLAND | 260* | Steam Yacht | June 1883 |
| 49 | SS CRAIGENDORAN | 1471 | Cargo Ship | Oct 1883 |
| 50 | SS CROSSHILL | 954 | Cargo Ship | Nov 1883 |
| 51 | HIGHLAND FOREST | 1040 | 3 masted steel barque | Mar 1884 |
| 52 | SY LADY NELL | 423* | Steam Yacht | Jan 1884 |
| 53 | SY MAZEPPA | 219* | Steam Yacht | Mar 1884 |
| 54 | PSS BOURNEMOUTH | 233 | Paddle Steamer | Apr 1884 |
| 55 | SY LADY ALINE | 338* | Steam Yacht | Apr 1884 |
| 56 | SS BORDEAUX | 550 | Cargo Ship | Aug 1884 |
| 57 | TS OTTER | 191 | Twin screw tug | July 1884 |
| 58 | SS MASCOTTE | 1100 | Cargo Ship | Jan 1885 |
| 59 | SS MALACCA | 653 | Cargo Ship | Feb 1885 |
| 60 | SY KATRENA | 268* | Steam Yacht | Apr 1885 |
| 61 | SS EL CALLAO | 1323 | Cargo Ship | Apr 1885 |
| 62 | CROWN OF INDIA | 2056 | 4 masted barque | June 1885 |
| 63 | SWPS HENRY VINN | - | Stern paddle wheel steamer | May 1885 |
| 64 | El Strivido | - | Steam Launch | May 1885 |
| 65 | SY LADY BEATRICE | 245* | Steam Yacht | Aug 1885 |
| 66 | CROWN OF ITALY | 1618 | 3 masted barque | Sept 1885 |
| 67 | HIGHLAND HOME | 1371 | 3 masted iron barque | Feb 1886 |
| 68 | SY VIOLET | 125* | Steam Yacht | Feb 1886 |
| 69 | SY MIRANDA | 250* | Steam Yacht | Apr 1886 |
| 70 | ST SOLWAY | 103 | Steam tug | Apr 1886 |
| 71 | SS BEAVER | 222 | Steam Ferry | Mar 1886 |

| Yard No | Name | Tonnage | Type | Launched |
|---------|------|---------|------|----------|
| 72 | CASTOR | 2059 | Barque | July 1886 |
| 73 | SS BAN SENG GUAN | 801 | Screw Steamer | July 1886 |
| 74 | SY GLADIATOR | 164* | Steam Yacht | July 1886 |
| 75 | SY RONDINE | 456* | Steam Yacht | Jan 1887 |
| 76 | SS FATSHAN | 2260 | Passenger Ferry | Mar 1887 |
| 77 | SY Little Violet | - | Steam Launch | Dec 1887 |
| 78 | SY DOTTEREL | 165* | Steam Yacht | Apr 1887 |
| 79 | SY MALIKAH | 222* | Steam Yacht | July 1887 |
| 80 | Atlantis | - | Yawl | July 1887 |
| 81 | TSS CHAMROEN | 418 | Twin screw Steamer | Dec 1887 |
| 82 | SS HAILOONG | 1253 | Cargo/Passenger Ferry | Feb 1888 |
| 83 | SY RED EAGLE | 204* | Steam Yacht | Mar 1888 |
| 84 | SY GARLAND | 179* | Steam Yacht | Apr 1888 |
| 85 | SS ANCONA | 1245 | Cargo Steamer | June 1888 |
| 86 | SS RAVENNA | 1243 | Screw Steamer | Aug 1888 |
| 87 | SS MARISTOW | 1679 | Screw Steamer | Dec 1888 |
| 88 | SS SUFFOLK | 3303 | Steel-Cargo Ship | Mar 1889 |
| 89 | SS ROSARY | 1162 | Cargo Steamer | Apr 1889 |
| 90 | SS REALM | 1712 | Cargo Steamer | July 1889 |
| 91 | SS REX | 1750 | Cargo Steamer | Aug 1889 |
| 92 | SS BARRACLOUGH | 1784 | Cargo Steamer | Sept 1889 |
| 93 | SY SEMIRAMIS | 491* | Steam Yacht | May 1889 |
| 94 | SS ZAMORA | 1245 | Cargo Steamer | Oct 1889 |
| 95 | SS WEIMAR | 1580 | Cargo Steamer | Dec 1889 |
| 96 | TSS HEUNG-SHAN | 1985 | Twin screw Steamer | Feb 1890 |
| 97 | SS TALUNE | 2087 | Cargo Steamer | Apr 1890 |
| 98 | ORION | 2080 | Barque | June 1890 |
| 99 | CAISSON | - | Barque | Dec 1889 |
| 100 | SY ST. GEORGE | 641* | Aux Steam Yacht | Aug 1890 |
| 101 | SS NEVADA | 1285 | Cargo Steamer | Aug 1890 |
| 102 | SY LADYE MABEL | 569* | Steam Yacht | Jan 1891 |
| 103 | SS TYNE | 615 | Cargo Steamer | Nov 1890 |
| 104 | TRADE WINDS | 2859 | 4-masted Barque | Feb 1891 |
| 105 | SS MOSCOW | 1622 | Cargo Steamer | Mar 1891 |
| 106 | SY ZARIA | 759* | Aux Steam Yacht | July 1891 |
| 107 | SY LADY INA | 311* | Steam Yacht | May 1891 |
| 108 | DRUMROCK | 3182 | 4-masted Barque | Aug 1891 |
| 109 | SS CORUNNA | 1269 | Cargo Steamer | Aug 1891 |
| 110 | WILHELM TELL | 3107 | 4-masted Barque | Nov 1891 |
| 111 | PROCYON | 2120 | 3-masted Barque | Dec 1891 |

| Yard No | Name | Tonnage | Type | Launched |
|---------|------|---------|------|----------|
| 112 | SS BERNICIA | 793 | Cargo Steamer | Feb 1892 |
| 113 | CROWN OF AUSTRIA | 3137 | 4-masted Barque | May 1892 |
| 114 | SY MAHA CHAKRKRI | 2229* | T.S. Cruiser Yacht | June 1892 |
| 115 | SY FAUVETTE | 330* | Steam Yacht | May 1892 |
| 116 | SS MINORCA | 1161 | Cargo Steamer | Aug 1892 |
| 117 | SY VALHALLA | 1211* | Aux Steam Yacht | Oct 1892 |
| 118 | SS NORSEMAN | 1117 | Cargo Steamer | Nov 1892 |
| 119 | SY GUNDREDA | 311* | Steam Yacht | Mar 1893 |
| 120 | ROYAL FORTH | 3130 | 4-masted Barque | May 1893 |
| 121 | SY CLEOPATRA | 544* | Steam Yacht | April 1893 |
| 122 | SS VALA | 1016 | Cargo Steamer | Dec 1893 |
| 123 | SY ERL KING | 444* | Steam Yacht | Mar 1894 |
| 124 | SS VANA | 1021 | Cargo Steamer | Feb 1894 |
| 125 | SY ZETA | 181* | Steam Yacht | Mar 1894 |
| 126 | SS NAROVA | 949 | Cargo Steamer | April 1894 |
| 127 | SY ST ELIAN | 203* | Steam Yacht | April 1894 |
| 128 | SY ELLIDA | 265* | Steam Yacht | May 1894 |
| 129 | SY LA BELLE SAUVAGE | 531* | Aux Steam Yacht | Oct 1894 |
| 130 | SS GEORGIOS | 698 | Cargo Steamer | Aug 1894 |
| 131 | SS SOFIA | 696 | Cargo Steamer | Sept 1894 |
| 132 | CARNAO | 117 | St Lighter | Sept 1894 |
| 133 | SS SAINT NINIAN | 717 | Cargo Steamer | Feb 1895 |
| 134 | SY HERSILIA | 330* | Steam Yacht | Mar 1895 |
| 135 | SY SPEEDY | 113* | Steam Yacht | Mar 1895 |
| 136 | SS NUBIA | 1196 | Scr Steamer | June 1895 |
| 137 | SY LADY SOPHIA | 203* | Steam Yacht | May 1895 |
| 138 | CORSICA | 1103 | Scr Steamer | July 1895 |
| 139 | Simanette | 20 | Steam Launch | 4 July 1895 |
| 140 | SY ARCTURUS | 432* | Aux Steam Yacht | Oct 1895 |
| 141 | SY IOLAIRE | 531* | Steam Yacht | Jan 1896 |
| 142 | SS ROSONA | 952 | Scr Steamer | Jan 1896 |
| 143 | TSS FRONTIER | 1191 | T.S. Steamer | May 1896 |
| 144 | SY SPEEDY II | 233* | T.S. St Yacht | April 1896 |
| 145 | SY Cooljack | | T.S. Steam Launch | Aug 1896 |
| 146 | SY ROSABELLE II | 292* | Steam Yacht | Feb 1897 |
| 147 | SS VESTRA | 1021 | Scr Steamer | Jan 1897 |
| 148 | SY LEON PAULIHAC | 215* | Steam Yacht | Feb 1897 |
| 149 | SY GUNILDA | 492* | Steam Yacht | April 1897 |
| 150 | SY Mandolin | | Steam Yacht | April 1897 |
| 151 | ST GAVIAO | 102 | St Tug | May 1897 |

| Yard No | Name | Tonnage | Type | Launched |
|---------|------|---------|------|----------|
| 152 | SY KETHAILES | 447* | Steam Yacht | May 1897 |
| 153 | SS RONAN | 1198 | Scr Steamer | Sept 1897 |
| 154 | Chicago | | Grain Elevator | Aug 1897 |
| 155 | SS LOUGA | 952 | Scr Steamer | Feb 1898 |
| 156 | SS CATHAY | 4112 | Scr Steamer | May 1898 |
| 157 | TSS NATUNA | 764 | T.S. Steamer | April 1898 |
| 158 | SS REVAL | 1682 | Scr Steamer | July 1898 |
| 159 | SY SURF I | 390* | Steam Yacht | Sept 1898 |
| 160 | SY SHEMARA | 477* | Steam Yacht | Feb 1899 |
| 161 | SS SIFKA | 1103 | Scr Steamer | Oct 1898 |
| 162 | SS ELBA | 1081 | Scr Steamer | May 1899 |
| 163 | SY LADYE GIPSEY | 373* | Steam Yacht | Jan 1899 |
| 164 | AURORA | 164 | Steam Trawler | Mar 1899 |
| 165 | ARIES | 159 | Steam Trawler | Mar 1899 |
| 166 | SY GOLDEN EAGLE | 356* | Steam Yacht | June 1899 |
| 167 | ST PANTHER | 236 | W.S.S. Tug | Nov 1899 |
| 168 | Barge | 158 | Barge | Sept 1899 |
| 169 | SS SAPPHO | 1275 | Scr Steamer | Jan-00 |
| 170 | ATLANTA | | Steam Trawler | Sept 1899 |
| 171 | SS PORT MARIA | 2910 | Scr Steamer | Oct-01 |
| 172 | SY BANSHEE | 993* | T.S. Steam Yacht | Jul-00 |
| 173 | SY ROSITA | 119* | Steam Yacht | 1900 |
| 174 | SS BENCLEUCH | 4159 | Scr Steamer | Nov-00 |
| 175 | TSS TRITON | 234 | T.S. Steamer | Mar-01 |
| 176 | ST HELEN PEELE | 133 | T.S. Tug | Jun-01 |
| 177 | SS BINTANG | 1404 | Scr Steamer | Jul-01 |
| 178 | SS RAJAH OF SARAWAK | 1452 | Scr Steamer | Sep-01 |
| 179 | SY ROSABELLE III | 526* | Steam Yacht | Nov-01 |
| 180 | SS BELRORIE | 210 | Salvage Steamer | Nov-01 |
| 181 | SS TAKESHAM MARU | 2012 | Scr Steamer | Feb-02 |
| 182 | MY LORENA | 1303 | Parsons Turbine Yacht | Jan-03 |
| 183 | New York | | Grain Elevator | Dec-01 |
| 184 | SY ARIANA | 285* | Steam Yacht | Mar-02 |
| 185 | SY RANNOCH | 479* | Steam Yacht | Apr-02 |
| 186 | SS SCALPA | 1010 | Scr Steamer | May-02 |
| 187 | SS STAFFA | 1008 | Scr Steamer | Aug-02 |
| 188 | SS VIENNA | 1912 | Scr Steamer | Jan-03 |
| 189 | SY WAKIVA I | 417* | Steam Yacht | Mar-03 |
| 190 | EVELYN | 131 | Schooner Yacht | Apr-03 |
| 191 | SS FARRALINE | 1226 | Scr Steamer | Jun-03 |

| Yard No | Name | Tonnage | Type | Launched |
|---------|------|---------|------|----------|
| 192 | SY ROVENSKA | 620* | Steam Yacht | Mar-04 |
| 193 | SS KUCHING | 1445 | Scr Steamer | Apr-04 |
| 194 | Miner 17 | | Submarine hunting Scr Steamer | Sep-04 |
| 195 | Miner 18 | | Submarine hunting Scr Steamer | Apr-04 |
| 196 | SY Einna | | Aux Steam Yacht | Jun-04 |
| 197 | SS JAMES CROMBIE | 626 | Scr Steamer | Sep-04 |
| 198 | SY HONOR | 1129* | T.S. Steam Yacht | Apr-05 |
| 199 | SS PLOUSSA | 987 | Scr Steamer | Apr-05 |
| 200 | SY VENETIA | 577* | Steam Yacht | May-05 |
| 201 | SS WRECKER | 451 | Scr Steamer | Aug-05 |
| 202 | SY MAUND | 903* | Steam Yacht | Jan-06 |
| 203 | SS MEYUN | 567 | Scr Steamer | Oct-05 |
| 204 | SY MINONA | 199* | Steam Yacht | Apr-06 |
| 205 | SY AGAWA | 602* | Steam Yacht | Sep-06 |
| 206 | SS MELROSE | 1653 | Scr Steamer | Aug-06 |
| 207 | SY O-WE-RA | 426* | Steam Yacht | Nov-06 |
| 208 | SS HUNTER | 1840 | T.S. Steamer | Feb-07 |
| 209 | SY WAKIVA II | 853* | T.S. Steam Yacht | Mar-07 |
| 210 | ST TERAWHITE | 260 | Tug | Apr-07 |
| 211 | SY LADY BLANCHE | 327* | Steam Yacht | Aug-07 |
| 212 | SS TUNA | 662 | Scr Steamer | Aug-07 |
| 213 | TS SY LIBERTY | 1607* | T.S. Steam Yacht | Dec-07 |
| 214 | TS SY IOLANDA | 1647 | T.S. Steam Yacht | Mar-08 |
| 215 | TSS KARUAH | 399 | T.S. Steamer | Jul-08 |
| 216 | SS ODER | 965 | Scr Steamer | Feb-09 |
| 217 | SS ALSTER | 964 | Scr Steamer | Mar-09 |
| 218 | SS SLEMISH | 1536 | Scr Steamer | Jun-09 |
| 219 | TSS ARGUS | 653 | T.S. Steamer | Oct-09 |
| 220 | SY LA RESOLUE | 696 | Aux Steam Yacht | Feb-10 |
| 221 | SS KINGSTOWN | 628 | Scr Steamer | Apr-10 |
| 222 | SS CARNDUFF | 257 | Scr Steamer | Jun-10 |
| 223 | SS ESTRELLANO | 1161 | Scr Steamer | Sep-10 |
| 224 | SY UL | 854 | Steam Yacht | May-11 |
| 225 | SS KANNA | 1948 | Scr Steamer | Apr-11 |
| 226 | SS SWIFT | 1141 | Scr Steamer | Aug-11 |
| 227 | TSS KOOPA | 416 | T.S. Steamer | Sep-11 |
| 228 | TSS ALERT | 777 | T.S. Steamer | Dec-11 |
| 229 | SS ST. MAGNUS | 957 | Scr Steamer | Mar-12 |

| Yard No | Name | Tonnage | Type | Launched |
|---------|------|---------|------|----------|
| 230 | SS KARUMA | 934 | Scr Steamer | Apr-12 |
| 231 | SS PRINCESS MELITA | 1094 | Scr Steamer | Aug-12 |
| 232 | SS BERNICIA | 957 | Scr Steamer | Oct-12 |
| 233 | SS PALMELLA | 1352 | Scr Steamer | Dec-12 |
| 234 | SS WENDY | 958 | Scr Steamer | Mar-13 |
| 235 | SS ST MARGARET | 943 | Scr Steamer | Apr-13 |
| 236 | SS THORNFIELD | 488 | Scr Steamer | Jun-13 |
| 237 | SS TRANSVAAL | 4395 | Scr Steamer | Feb-14 |
| 238 | SS CHAKDARA | 3055 | Scr Steamer | Jun-14 |
| 239 | SS CHAKDINA | 3033 | Scr Steamer | Sep-14 |
| 240 | SS WESTMORELAND | 1765 | Scr Steamer | Jan-15 |
| 241 | SS BELTOY | 1544 | Scr Steamer | Apr-15 |
| 242 | Black Dragon | 3400 | Hulk | 1916 |
| 243 | SS DARINO | 1433 | Scr Steamer | Jul-17 |
| 244 | SS SERBINO | 4080 | Scr Steamer | Mar-19 |
| 245 | X-93 | 130 | Landing Craft | May-15 |
| 246 | X-94 | 130 | Landing Craft | Apr-15 |
| 247 | X-95 | 130 | Landing Craft | Apr-15 |
| 248 | X-218 | 120 | Landing Craft | Dec-16 |
| 249 | X-219 | 120 | Landing Craft | Jan-17 |
| 250 | INVER | 1542 | Scr Steamer | Jun-19 |
| 251 | ALERT | | Scr Steamer | Oct-19 |
| 252 | X-96 | 120 | Landing Craft | May-15 |
| 253 | SS MATAMA | 2012 | Scr Steamer | Dec-19 |
| 254 | SS PALMELLA | 1568 | Scr Steamer | Apr-20 |
| 256 | SS RUNO | 1858 | Scr Steamer | Oct-20 |
| 257 | KØBENHAVN | 3901 | 5 masted sailing ship | Mar-21 |
| 258 | SS TASSO | 3540 | Scr Steamer | Jun-22 |
| 259 | SS KARA | 1846 | Scr Steamer | Oct-21 |
| 260 | SS MALVERNIAN | 4373 | Scr Steamer | Jun-24 |
| 261 | SS HARROGATE | 1029 | Scr Steamer | Oct-24 |
| 262 | TSMY NAZ-PERWER | 598 | T.S.M. Yacht | Dec-23 |
| 263 | TSMY SEABORN | 510 | T.S.M. Yacht | Jul-25 |
| 264 | TSMY EROS | 1112 | T.S.M. Yacht | Mar-26 |
| 265 | SS VYNER BROOKE | 1670 | Scr Steamer | Nov-27 |
| 266 | SS OPORTO | 2352 | Scr Steamer | May-28 |
| 267 | Barge | - | Barge | - |
| 268 | SS HORSA | 979 | Scr Steamer | Aug-28 |
| 269 | SY NAZ-PERWER | 708 | Steam Yacht | Sep-30 |
| 270 | MERCATOR | 770 | 3 masted barquentine | Dec-31 |

We can see the form of the ship taking place in this undated photograph from the Leith shipyard of Henry Robb. It shows a ship in frame in the early 1930s; until the introduction of welding at Leith in the early 1960s this was the traditional way of building ships. The surrounding and underfoot conditions never changed much, right up to the closure of the last shipyard at Leith in 1984. (Author's collection)

# GLOSSARY

Shipbuilding has developed its own language, with thousands of words which have been coined over time to explain the particulars of working on the build of a vessel or when working on the actual vessel. Some are only used in boatbuilding while some are specific to steel shipbuilding; others are specific to either segment of what is a vast and complex industry.

The lists below explain the abbreviations and terms used in this book. NB terms in **bold** can be found as headwords in the Terms list.

## ABBREVIATIONS

| ABV | armed boarding vessel |
|-----|----------------------|
| aux | auxiliary motor |
| CS | cable ship |
| DAMS | defensively armed merchant ship |
| DEMS | defensively equipped merchant ship |
| DWT | **deadweight tons** |
| FV | fishing vessel |
| grt | gross registered tonnage |
| HMAS | His Majesty's Australian ship |
| HMNZT | His Majesty's New Zealand transport |
| HMS | His Majesty's ship |
| HMT | His Majesty's trawler |
| ihp | indicated horse power |
| LOA | length overall; the maximum length of the vessel |
| LWL | length at waterline |
| MT | motor tug |
| M/T | measurement ton (see under **Ton**) |
| MV | motor vessel |

| ON | official number |
|---|---|
| PS | paddle steamer |
| PSS | paddle steam ship |
| PTB | patrol boat |
| RFA | royal fleet auxiliary |
| RMS | Royal Mail ship |
| RNLI | Royal National Lifeboat Institute |
| Scr | screw-propelled; the vessel has a shaft driving a propeller (as against paddle-wheel propulsion) |
| SHP | shaft horsepower |
| SMS | *Seiner Majestät Schiff* (His Majesty's ship) |
| SS | steam ship |
| ST | steam tug |
| SWPS | stern-wheel paddle steamer |
| SY | steam yacht |
| THV | Trinity House vessel |
| TM | Thames measurement |
| TS | twin screw |
| TSM | twin-screw motor |
| TSMV | twin-screw motor vessel |
| TSMY | twin-screw motor yacht |
| TSS | twin-screw steamer |
| TSSY | twin-screw steam yacht |
| U-boat | *unterseeboot* (under-sea boat), identified by the letters UB, then the given number |
| USS | United States ship |
| W/T | weight ton (see under **Ton**) |

## Shipbuilding and nautical terms

| **ABV** | armed boarding vessel |
|---|---|
| **Anchor** | A heavy, pick-like device attached to a boat's stem by a **warp** and chain. Modern anchors are made of steel; common types of anchor are plough, fisherman and Danforth. The chain, which connects the anchor to its warp, is fixed onto the lower anchor end, adding weight and preventing chafing of the warp on rocks or shellfish beds. |
| **AP (After perpendicular)** | A line vertical from the baseline, usually taken from the after side of the rudder stock, or sometimes through the centre of the rudder stock. This line is taken up to where it intersects with the **load** or reference **waterline**. From the point where that load waterline passes through the ship's stem is created the FP or Forward Perpendicular. |
| | The distance between the two lines, known as length between perpendiculars or LBP, is divided up by the naval architect to create what is known as frame stations. They will form the grid whereby the naval architect will begin to create the small-scale lines for the ship known as scantling lines. (see also under **loftsman**). |

| | |
|---|---|
| **Base line** | The design line that all forward and aft measurements are taken, from the **AP** or the **FP**. All height measurements are taken from the baseline; in the case of a **drop keel** these heights will be shown as a minus. These measurements are used to create part of the **offset table**. |
| **Beam** | The maximum width of the ship. |
| **Beam knee** | A bracket holding the transverse deck beam to the vertical frames at the side of the hull. |
| **Bilge** | The lowest part of the hull interior, under the **sole**. Water and or fuel tanks are often placed in the bilges to lower the centre of gravity and so help keep the ship upright. |
| **Bilge keel** | A longitudinal, external, underwater member used to reduce a ship's tendency to roll and to aid directional stability. In Britain twin bilge keels are often used on small boats moored in estuaries with a large tidal range so the boat stays upright when the mooring dries out. With their much shallower draft, yachts of this type can be sailed in shallow waters. Not as hydrodynamically efficient as a fin keel. |
| **Bilge pump** | A pump, either manual or electric, with the inlet set at the lowest point in the bilges, where water will collect when the boat is upright. The inlet is protected by a screen to stop blockages. |
| **Black Squad** | The collective name for all the steel-working trades that built a ship. |
| **Block** | See **pulley**. |
| **Bow** | The front and generally sharp end of the hull. It is designed to reduce the resistance of the hull cutting through water and should be tall enough to prevent water from easily washing over the deck of the hull. The bulbous bow designed into some larger ships since the 1920s improves speed and stability. |
| **Bowsprit** | On a sailing ship, a spar that extends forward from the foredeck, outboard of the hull proper. Common in square-rigged ships, where they were used to attach the outer or flying jib. In modern sailing boats they are often made of lightweight carbon and are used to attach the **luff** of lightweight sails such as spinnakers. |
| **Breast hook** | The brackets that hold the shape of the **soft nose**. |
| **Bridge** | (*in ship***building**) A means of temporarily connecting two plates together to be **faired** and welded. |
| **Bulkhead** | The internal transverse structure of the hull; the number of bulkheads is determined by the length of the ship under class rules. |
| **Bulwark** | The upstanding part of the topsides around the edges of the deck, providing some security when a boat is heeled. |
| **Capstan** | A vertical metal or wooden winch secured to the foredeck of a ship, used for hoisting the anchor. Capstans may be manually operated, or powered hydraulically or electrically. A traditional sailor-powered wooden capstan is fitted with removable spoke-like wooden arms which the sailors push round and round, often in time to a sea shanty or chant. |
| **Coffin plate** | The plate joining two side plates over the **keel** of a vessel at the **stern**, which in plan view creates a shape similar to a coffin lid. |
| **Complement** | The full number of people required to operate a ship. Includes officers and crew members; does not include passengers. The number of people assigned to a warship in peacetime may be considerably less than her full complement. |
| **Counter stern** | A traditional stern construction with a long overhang and a shorter, upright, end piece. The counter is usually decked over. The stern is rounded when seen in plan view; other shapes of stern are **transom**, elliptical and round. |

| Cube | The cargo-carrying capacity of a ship, measured in cubic feet. There are two common types: | |
|---|---|---|
| | **Bale Cube** (or **Bale Capacity**) | A measurement of capacity for cargo in bales, on pallets etc., where the cargo does not conform to the shape of the ship. The space available for cargo is measured in cubic feet to the inside of the cargo battens on the frames, and to the underside of the beams. |
| | **Grain Cube** (or **Grain Capacity**) | A measurement of capacity for cargo like grain, which flows to conform to the shape of the ship. The maximum space available for cargo is measured in cubic feet, the measurement being taken to the inside of the shell plating of the ship or to the outside of the frames, and to the top of the beam or underside of the deck plating. |
| **Deadrise** | The difference in height between the **base line** and the point where a straight line through the flat of bottom surface intersects a vertical line through the side of the moulded surface at its widest point transversely | |
| **Deck** | The top surface of the hull, which keeps water and weather out of the hull and allows the crew, standing and walking on it, to operate the boat more easily. It stiffens the hull. Temporary frames (or moulds) can be removed and kept for another boat. | |
| **Displacement** | A measurement of the weight of the vessel, usually used for warships. Merchant ships are usually measured based on the volume of cargo space; see **Tonnage.**) Displacement is expressed either in long tons of 2,240 lbs, or in metric tonnes of 1,000 kg. Since the two units are very close in size (2,240 pounds = 1,016 kg and 1,000 kg = 2,205 lbs), it is common not to distinguish between them. To preserve secrecy, nations sometimes mis-state a warship's displacement. | |
| | **Displacement, light** | The weight of the ship excluding cargo, fuel, ballast, stores, passengers and crew, but with water in the boilers to steaming level. |
| | **Displacement, loaded** | The weight of the ship including cargo, passengers, fuel, water, stores, dunnage and such other items necessary for use on a voyage, which brings the vessel down to her load draft. |
| | **Deadweight tons** (DWT) | A measure of the ship's total carrying capacity. It's the difference between **displacement, light** and **displacement, loaded**. |
| | **Cargo deadweight tons** | The weight remaining after deducting fuel, water, stores, dunnage, and other such items necessary for use on a voyage, from the deadweight of the vessel. |
| **Dog** | A means of holding down iron or steel plate. The more dogs that are required the more distortion found in a plate; this is usually due to poor welding sequence, and a sign of re-work needed, hence many more man-hours to do a job that has not been dimensionally controlled. | |
| **Draught** | Distance from the bottom of the keel to the waterline. It may alternatively be spelt 'draft'. | |
| **Draught, loaded** | The depth of water necessary to float a vessel fully loaded. | |
| **Drop keel** | A retractable or removable fin / centreboard / daggerboard. | |
| **Fairing** | (*in shipbuilding*) ensuring that material put in place is correctly located | |
| **Forward perpendicular (FP)** | See **After perpendicular**. | |

| | |
|---|---|
| **Frame** | The transverse structure that gives a boat its cross-sectional shape. Frames may be solid or peripheral. They may be made of wood, plywood, steel, aluminium or composite materials. They may be removed after construction to save weight, or be reused, or left *in situ*. |
| | In ancient shipbuilding the frames were put in after the planking, but now most boats are built with the frames first. This gives greater control over the shape. |
| | In old, heavily built square-rigged ships, the frames were made up of four individual timbers, futtocks, as it was impossible to make the shape from a single piece of wood. The futtock closest to the keel was the ground futtock and the other pieces were called upper futtocks. |
| **Freeboard** | The distance between the waterline and the **deck** when loaded. Boats using sheltered waters can have low freeboard, but seagoing vessels need high freeboard. |
| **Freeboard, moulded** | The difference between the moulded depth and the moulded draft. (It is the height of the side of the vessel which is above the water when she floats at her **load waterline**.) |
| **Graving dock** | Dry dock. |
| **Gunwale** | The upper, outside longitudinal structural member of the **hull**. |
| **Hatch** | A lifting or sliding opening into the cabin, or through the deck for loading and unloading cargo. |
| **Heads** | Marine toilet. An abbreviation of the term 'catheads', which, up at the bow, were the normal place for toileting in square-rigger days. Always used in the plural. (The designed function of the catheads, timbers set outboard of the hull, was to provide protection for the hull against the friction of the anchor warp.) |
| **Howff** | The inner sanctum within some shipyards, usually trade-specific, so the shipwrights in their various groups would have howfs scattered around the shipyard: the platers would have their own howf, as would the welders, and so on. |
| **Hull** | The main body of a ship or boat, including her bottom, sides and deck. Some people are surprised to find that in a modern ship of any size the hull, for most of its length, has a flat bottom the full width of the ship. This is, however, essential for stability, because when such a ship heels then on its lower side there is a greater air-filled volume under the water than on the upper side, thus pushing the ship back upright, countering the heel. |
| **Iron Fighters** | A colloquial west of Scotland / Clyde shipbuilding term for the **Black Squad**. |
| **Keel** | The main central member along the length of the bottom of the ship or boat. It is an important part of the ship's structure, which also has a strong influence on its turning performance and in sailing ships resists the sideways pressure of the wind, enabling the course to be steered. |
| **Keelson** | An internal beam fixed to the top of the **keel** to strengthen the joint of the upper members of the ship to the keel. |
| **Kevlar** | A hi-tech and very strong synthetic material, used for cables, bulletproof jackets etc, developed by DuPont in the 1960s. |
| **Knuckle** | Where a plate changes angle or direction (creates effectively a fold in a plate); a good way of reducing the amount of welding which is required. This method of platework also reduces distortion when carried out properly. |
| **Length** | The distance between the forward-most and aftermost parts of the ship. |

| | | |
|---|---|---|
| | *Length overall* (LOA) | The maximum length of the ship. |
| | *Length when submerged* (LOS) | The maximum length of the submerged hull measured parallel to the designed **load waterline**. |
| | *Length at waterline* (LWL) | The ship's length measured at the waterline. |

| Load waterline | A line created during the initial design phase of a ship; it starts as a reference waterline showing where the ship can be loaded up to. |
| | During the ship's life the load waterline may be changed, but the original design reference waterline will always be so. |
| **Lofting** | The process used to create life-size drawings of **frames** so they can be manufactured. Today frames can be cut by a robot directly from a computer programme with extreme accuracy. |
| **Loftsman** | The loftsman (loft) was responsible for taking the scaled-down naval architect's scantling lines and producing them full size or one-tenth scale so the ship's form could be made. |
| | The loft then took the finished faired offsets from this and developed the shell plating and all the other steelwork involved in building the vessel. Templates were made so that the curved forms could be cut out of steel full size, along with the dimensional control of the build. |
| | Today this is all done by Computer Aided Design (CAD) generating a 3D model from which the ship's production drawings are taken. The role of the loft has been replaced by the designer and the nesting team, the dimensional control team, and quality control and planning. |
| **Luff** | The front part of a fore-and-aft sail, attached to either the **mast** or a **stay**. |
| **Mast** | A vertical pole on a ship which supports sails and/or rigging. If it's a wooden multi-part mast, this term applies specifically to the lowest portion. |
| **Mast step** | A socket, often strengthened, to take the downward thrust of the mast and hold its foot in position. In smaller craft, the mast step may be on the keel or on the deck. |
| **Mizzen** | In a sailing boat with two or more masts, a permanent mast and sail set aft of the mainmast. |
| **Moulded line** | Where all design measurements start from and are measured to. |
| **Offset table** | Used in ship design; contains measurements that give the coordinates for the lines plan (showing the curved lines that indicate the shape of the hull). |
| **O-Gee** | The line of the shaped plate that forms the 3D line from the main deck to the forecastle deck at the ship's side. |
| **Planing** | When the bow of the ship or boat, moving rapidly, lifts clear of the water. This is more hydrodynamically efficient, so is designed into speedy vessels. |
| **Port** | The left side of the ship when looking forward; so called by the Vikings as this was the side that would go alongside a harbour wall. The opposite side of the vessel to **Starboard**. |
| **pratique** | The licence given to a ship to enter the port on assurance from the captain to the authorities that she is free from contagious disease. The clearance granted is commonly referred to as Free Pratique. |
| **Pulley** | A small, part-enclosed wheel used to help redirect the angle of a rope or, in combination with more pulleys set up as a block and tackle, to reduce the power needed to pull the object controlled by the rope. |
| **Rigging (standing)** | Wires rods, cables or ropes used to keep a mast upright. Since the 1960s stainless steel wire has become universal in the developed world. Elsewhere galvanised wire or even rope may be used because of its availability and cheapness. |
| | The type of stainless steel wire commonly used in standing rigging such as stays is Type 1 × 19: a non-flexible wire. |
| | The common way of attaching wire is to form a small loop at the end which is fixed in place by clamping a soft metal swag over the free ends. (Talurite is a common brand of swagging.) The wire loop is then fastened to a rigging screw, with a bow shackle to the chain plate. In small sailing boats Kevlar rope is sometimes used in place of wire. |

| | |
|---|---|
| **Rigging (running)** | The ropes or cables used in sailing ships to control the sails. Cables are of two types: |
| | • Type 7 × 7: a semi-flexible wire used for luff wires in sails, halyards (sometimes plastic-coated) trapeze wires and light halyards. |
| | • Type 7 × 19, which is used for all halyards, wire sheets, vangs and strops that must run through a pulley. |
| **Scuppers** | Gaps in the bulwarks which enable sea or rainwater to flow off the deck. |
| **Shackle** | A quick-release metal device used to connect cables to fixings. |
| **Shaft horsepower** (SHP) | The amount of mechanical power delivered by the engine to a propeller shaft. In the SI system of units one horsepower is equivalent to 746 watts. |
| **Sheave** | A **pulley**. |
| **Sheer** | The generally curved shape of the top of the **hull** when viewed in profile. The sheer is traditionally lowest amidships, to maximize **freeboard** at the ends of the hull. |
| | Sheer can also be reverse – higher in the middle to maximise space inside – or straight, or a combination of shapes. |
| **Sheet** | A rope used to control the position of a sail; e.g. the main sheet controls the position of the main sail. |
| **Skeg** | A long, tapering piece of timber fixed to the underside of a **keel** near the **stern** in a small boat, especially a kayak or rowing boat, to aid directional stability. |
| **Soft nose** | The upper strakes of plate that form the bow of the ship, extending from the solid stem bar to the main deck. |
| **Spar** | A length of timber, aluminium, steel or carbon fibre of approximately round or pear-shaped section, used to support sails; types of spar include a **mast**, boom, gaff, yard, **bowsprit**, prod, boomkin, pole and dolphin striker. |
| **Spring** | The amount of curvature in the keel from **bow** to **stern** when viewed side on. The modern trend is to have less spring (known as hogging or sagging) in order to have less disturbance to water flow at higher speeds, thus aiding **planing**. |
| **Starboard** | The right side of the ship when looking forward. The word comes from the Viking, whose ships had a *styrbord*, steering oar, on that side. The opposite side of the vessel to **Port**. |
| **Stays** | See **Rigging (standing)**. |
| **Steamer** | Steam-powered cargo or passenger ship. |
| **Stem** | A continuation of the **keel** upwards at the front of the **hull**. |
| **Stern** | The back of the boat. |
| **Strake** | A strip of material running longitudinally along the vessel's side, bilge or bottom. On a steel boat each longitudinal strake of plating has a name, such as Garboard strake, Bilge strake and Sheer strake, with any strakes of plating in between labelled A, B, C etc. |
| **Strong back** | A heavy plate used as a **fairing** aid to keep plates straight and fair, and to assist in the alignment before and during welding. |
| **Swag** | See **Rigging (standing)**. |
| **Taff rail** | A railing, often ornate, at the extreme stern of a traditional square-rigged ship. In light air conditions an extra sail would be set on a temporary mast from the taff rail. |
| **Ton** | The unit of measure often used in specifying the size of a ship. There are three completely unrelated definitions for the word; two of them refer to volume (the word was originally 'tun', a large barrel) and the third definition relates to weight. |

| | | |
|---|---|---|
| | *Measurement ton* (M/T) or *Ship ton* | Calculated as 40 cubic feet of cargo space. See **Cube: Bale Cube.** For example, a vessel with a capacity of 10,000 M/T has a bale cubic of 400,000 cubic ft. |
| | *Register ton* | A measurement of cargo-carrying capacity in cubic feet. One register ton is equivalent to 100 cubic feet of cargo space. |
| | *Weight ton (W/T)* | Calculated as a long ton (2,240 lbs). |
| **Tonnage** | A measurement of the cargo-carrying capacity of merchant vessels. It depends not on weight but on the volume available for carrying cargo. The basic units of measure are the Register Ton, equivalent to 100 cubic feet, and the Measurement Ton, equivalent to 40 cubic feet. The calculation of tonnage is complicated by many technical factors. | |
| | *Gross tons* | The entire internal cubic capacity of the ship expressed in tons of 100 cubic feet to the ton, except certain spaces which are exempted such as: peak and other tanks for water ballast, open forecastle bridge and poop, access of hatchways, certain light and air spaces, domes of skylights, condenser, anchor gear, steering gear, wheel house, galley and cabin for passengers. |
| | *Net tons* | Obtained from the gross tonnage by deducting crew and navigating spaces and making allowances for propulsion machinery. |
| **Transom** | A wide, flat or slightly curved, sometimes vertical, board at the rear of the hull, which on small power boats is often designed to carry an outboard motor.<br><br>Transoms increase width and buoyancy at the stern. On a boat designed to be powered by an outboard, the stern is often the widest point, to provide displacement to carry the heavy outboard and to resist the initial downward thrust of the craft when it's **planing**. | |
| **Warp** | A rope normally used for holding a vessel in place, either alongside a quay or another vessel, or to a buoy or anchor. | |